Visual Hydrology

A primer for interactive computing

Visual Hydrology

A primer for interactive computing

Peter A Mantz

DIC, PhD, CEng, CPhys

Visiting Professor, University of Newcastle, UK

IWA Publishing

Published by IWA Publishing, Alliance House, 12 Caxton Street, London SW1H 0QS, UK
Telephone: +44 (0) 20 7654 5500; Fax: +44 (0) 20 7654 5555; Email: publications@iwap.co.uk
Web: www.iwapublishing.com

First published 2004

British Library Cataloguing in Publication Data
A CIP catalogue record for this book is available from the British Library

Library of Congress Cataloging- in-Publication Data
A catalog record for this book is available from the Library of Congress

ISBN: 1 84339 056 6

CONTENTS

PREFACE

Hydrology attempts to quantify the mass of water in its movement from precipitation (as snow and rainfall), through distribution (as runoff and streamflow) and consumption (via reservoirs), and finally back to the environment (as wastewater). This "Visual Hydrology" Primer demonstrates the use of modern Information Technology towards presenting the subject. It combines theory with computational exercise, and is supplemented with an explanatory text for each code. The computer platform used is the commonly available "Microsoft Excel " software that includes the "Visual Basic for Applications" code.

A possible approach for introducing computational material to the classroom is to supplement a textbook with modern computer codes. Unfortunately most codes are expensive, designed for commercial use, without source code and may require special software. There is a need for a new layer of cheaper and simpler computational exercise that can be fully assimilated by a student. This layer should involve the activation, understanding and reproduction of modern computer code. "Visual Hydrology" attempts to meet this need by applying the following objectives:

- explaining the structure of modern object-oriented computer code;
- providing the source code for worked examples;
- numerically checking the worked examples used in the text;
- detailing how the worked examples can be used with alternative data;
- describing and referencing the underlying theory;
- providing additional exercises with each worked example;
- using Microsoft Excel 2000 or 2002 software alone.

This Primer should be of interest to students, research workers and consulting engineers in the water related sector. It is not a comprehensive exposition of all hydrological worked examples. It follows the practice of past monographs (for example, "Basic Hydrology", 1984, Butterworths Publishing Company), by demonstrating codes for common and practical examples used in hydrological engineering. The main purpose is to illustrate and teach the use of modern and readily available computer code towards engineering computation. A basic knowledge of Microsoft Excel is required as a background to the text, and several introductory references are given in the Computing section of the Bibliography.

The Visual Basic object model is introduced in the first chapter, with the aid of 5 worked examples that are called "Projects". Since the understanding and programming of these 5 introductory projects provide the necessary background for the remainder of the primer, the code is fully described in the text. The source codes for all the projects in the primer are electronically provided, with the exception of this first chapter. A reason for the latter is that the purpose of this primer is to involve the reader in learning the basics of interactive computer programming, and this must be achieved from the beginning chapter. (Note that this code is copyrighted, and network use is prohibited unless a separate license is acquired for each machine of intended use.)

Chapter 2 prepares the reader with the necessary background of statistics that are used in hydrology. The manipulation of matrices is introduced, and it soon becomes apparent why the computer is necessary for such operations. The modern topics of Linear moment analysis and Geostatistics are also presented. The geostatistical projects deal fully with Point and Block Kriging, and the code may be readily extended to other forms of Kriging (for example, Co-Kriging). These fundamental statistical topics are applied to hydrological subjects in subsequent chapters.

The analysis of groundwater flow is dealt with in Chapter 3. The first project, "Groundwater levels", demonstrates the use of Kriging in hydrology. The remaining 2 projects are applications of viscous flow theory to groundwater (Darcy's Law). The second project computes the drawdown when groundwater is consumed. And the third project introduces the application of the finite difference method to well drawdown.

Chapter 4 again demonstrates the use of Kriging, but this time for precipitation examples. The first project uses Point Kriging for designing the location of rainfall gauges, and the second project uses Block Kriging for minimising the error involved when calculating the mean rainfalls for a catchment. The FEH (UK Flood Estimation Handbook, 1999) method for computing rainfall depth, duration and frequency is illustrated in the third project.

The runoff of precipitation from ground surfaces to channels is dealt with in Chapter 5. Runoff is usually measured in rivers and streams by velocity traverse, or by some form of hydraulic structure. There are very many designs for the latter, so the commonly used horizontal broad-crested weir is chosen for example. The detailed analysis of weirs requires the knowledge of boundary-layer fluid mechanics, and this is briefly introduced in the text. This chapter also demonstrates the FEH method for determining the risk frequency of annual maximum flows.

Although it is usual for hydrological texts to begin with a discussion of the hydrological cycle, this subject is left to Chapter 6. The reason is that the 2 projects presented in this chapter compute a design hydrograph of river flows for a particular rainfall hyetograph. The preceeding chapters on precipitation and runoff are therefore necessary for sequential understanding. The 2 projects in this chapter contrast the UK FEH method and the US SCS (Soil Conservation Service) method for computing rainfall and runoff for a catchment. The latter project introduces the use of a "Knowledge Based Expert System" in hydrology.

Finally, Chapter 7 deals with the subject of reservoirs. The use of reservoirs in attenuating a flood hydrograph is computed in the first project using a finite difference approximation. And the calculation of reservoir storage capacity for enabling a constant yield flow is estimated in the second project.

The text used in this primer has 4 sources. These are the Excel toolbar commands (otherwise known as menus), Visual Basic objects, Visual Basic code, and the Author's names for VB projects, modules and code. For clarity, these sources use different punctuation and fonts as follows:

1. Excel toolbars and toolbar commands are enclosed with square brackets, e.g. [Command].
2. VB objects have the first letter capitalised and use the Arial 12 point font, e.g. Spin button.
3. VB code uses the Arial 12 pt. font.
4. Author's names have the first letter capitalised and are enclosed with inverted commas, e.g. "Draw.xls".

The codes for these worked examples were prepared using monitor display settings of 1024 * 768 pixels with Normal Size (96 DPI) fonts. This appears to be the current standard display setting for 14 inch, TFT monitors. At least 6 of the 23 projects use moving objects that require the programming of worksheet drawing units. Since the latter change with the screen font size, the display settings must be set to Normal size fonts for these 6 projects to be drawn correctly. Furthermore, for display settings less than 1024 * 768 pixels, only a part of some drawings will be visible using a full screen. It is then necessary to use the Excel [View.Zoom] command to view the entire drawing. At [Zoom] values less than 50%, the drawing becomes badly distorted. The projects are however usable for a minimum display setting of 800 * 600 pixels with Normal size fonts.

This preface must include an apology towards presentation. Firstly, the abbreviation of variables used in both the text and the computer codes are sometimes dissimilar, since code variables cannot accept parentheses or variable upper and lower case. The occasional double representation of variables is therefore unavoidable. Secondly, although every code has been tested using worked examples from many textbooks, there will be errors and omissions. The author therefore welcomes any positive criticism.

Finally, although the data for most exercises have been randomly generated where possible, my thanks are due to the following authorities who have generously supplied data examples:

Clark, I and Harper, W.V. (2000) "Practical Geostatistics 2000", Ecosse North America LLC, USA, http://uk.geocities.com/drisobelclark/briefcase.html
CEH Wallingford, Centre of Ecology and Hydrology, http://www.nwl.ac.uk/feh
UK Meteorological Office, Bracknell, Berkshire, http://www.metoffice.com

Peter A Mantz, July 2003
Visiting Professor
Water Resources Systems Research Laboratory
University of Newcastle
Newcastle Upon Tyne NE1 7RU
p.a.mantz@ncl.ac.uk

FOREWORD

The teaching of science and engineering subjects has traditionally been supported by laboratory experimentation. Students then have the opportunity to conduct experiments that develop their understanding of theoretical concepts which are developed in lectures. However, the extent to which laboratory experimentation can be employed within contemporary undergraduate and postgraduate curricula is limited by a number of factors. Firstly, many theoretical concepts do not easily lend themselves to laboratory experimentation. Secondly, pressure on university department budgets has led to both reductions in the number of skilled laboratory technicians who maintain laboratory facilities, and in the number of academic staff who deliver undergraduate and postgraduate degree programmes. There is currently a strong drive for improved efficiency in the delivery of residential programmes, and the recent development of distance learning programmes has created a need for new modes of delivery which can enhance the quality of the learning experience. Moreover, students pursuing distance (or flexible) learning programmes do not have the same opportunity as residential students for undertaking laboratory experimentation.

Since the arrival of the Desktop Computer, the traditional science or engineering textbook has frequently benefited from having an accompanying program disk to supplement the material. In 'Visual Hydrology', the author has taken this support to a new level by creating a Virtual Teaching Laboratory (VTL) for Hydrology where experimentation can be conducted using interactive computing. 'Visual Hydrology' is conceived as a primer in interactive computing; it demonstrates how the VTL concept can be implemented successfully in the field of hydrology. VTL modules have been developed for a number of statistical and process-based methods of hydrological analysis, using object-oriented code that is located in the commonly available 'Microsoft Excel' software. Since most students are familiar with this software, they can readily gain an understanding of how VTL code has been constructed, and indeed experiment with the development of simple modules themselves. VTL software modules are designed to allow the students to experiment with different parameters and design variables, and to view the results using moving graphics (as opposed to static graphics in textbooks).

The Visual Hydrology concept arose as a result of the author's enthusiasm for enhancing the student learning experience for the postgraduate distance learning programme in 'Hydroinformatics and Management Systems' (http://www.ncl.ac.uk/informatics). This program has been recently developed at the University of Newcastle. By setting up VTLs to be web-based, a significant advance in learning potential for off-campus students can be created, while residential students can also benefit from access to VTLs. The latter augment or supplement more traditional tutorial sessions for which individual tutoring is frequently not possible. Indeed, difficult technical concepts that are traditionally explained in lectures can be communicated more effectively through the use of VTLs.

I am pleased to have this opportunity to endorse the 'Visual Hydrology' approach, and to commend it to all students of hydrology. It should enrich their learning experience.

Professor Enda O'Connell
Water Resource Systems Research Laboratory
School of Civil Engineering and Geosciences
University of Newcastle upon Tyne

SOURCE CODE EXAMPLES

Examples of source code to accompany this publication can be found at:

http://www.iwapublishing.com/template.cfm?name=isbn1843390566

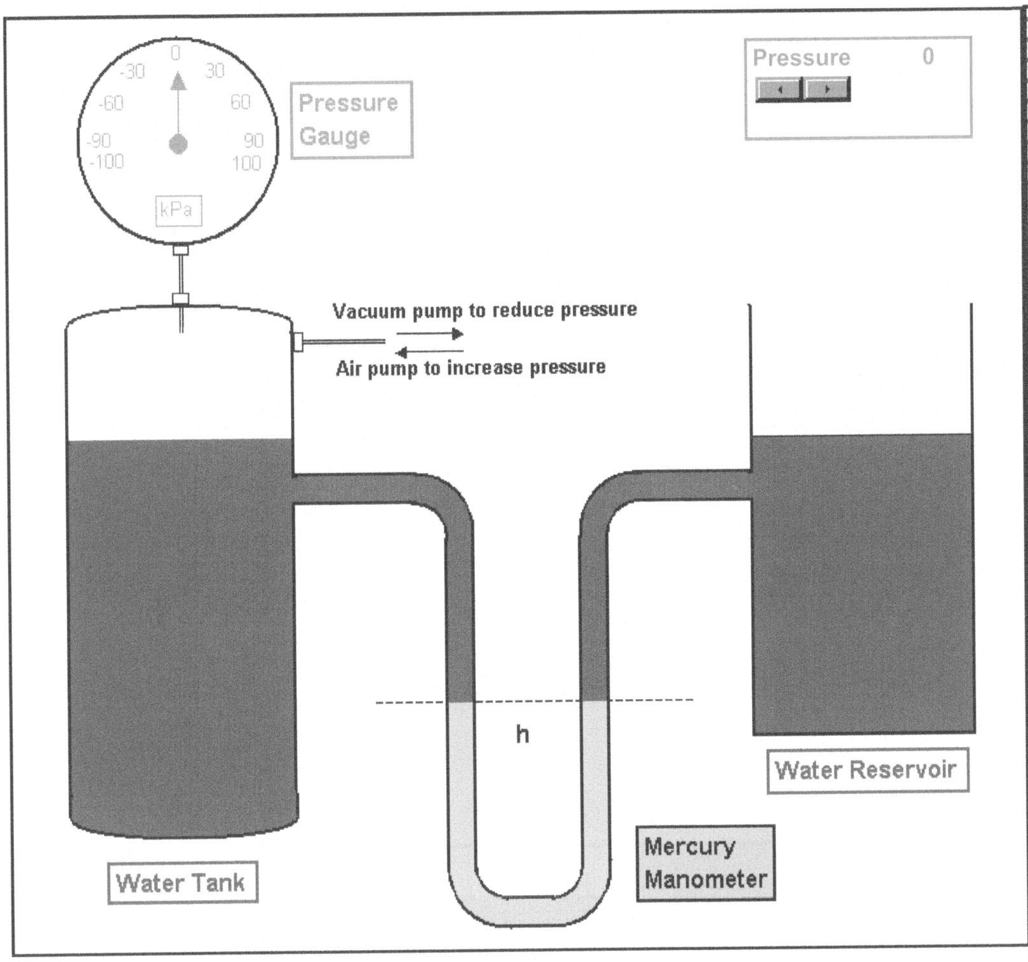

1 Visual Basic for Applications

1.1 THE VISUAL BASIC OBJECT MODEL

1.1.1 Summary

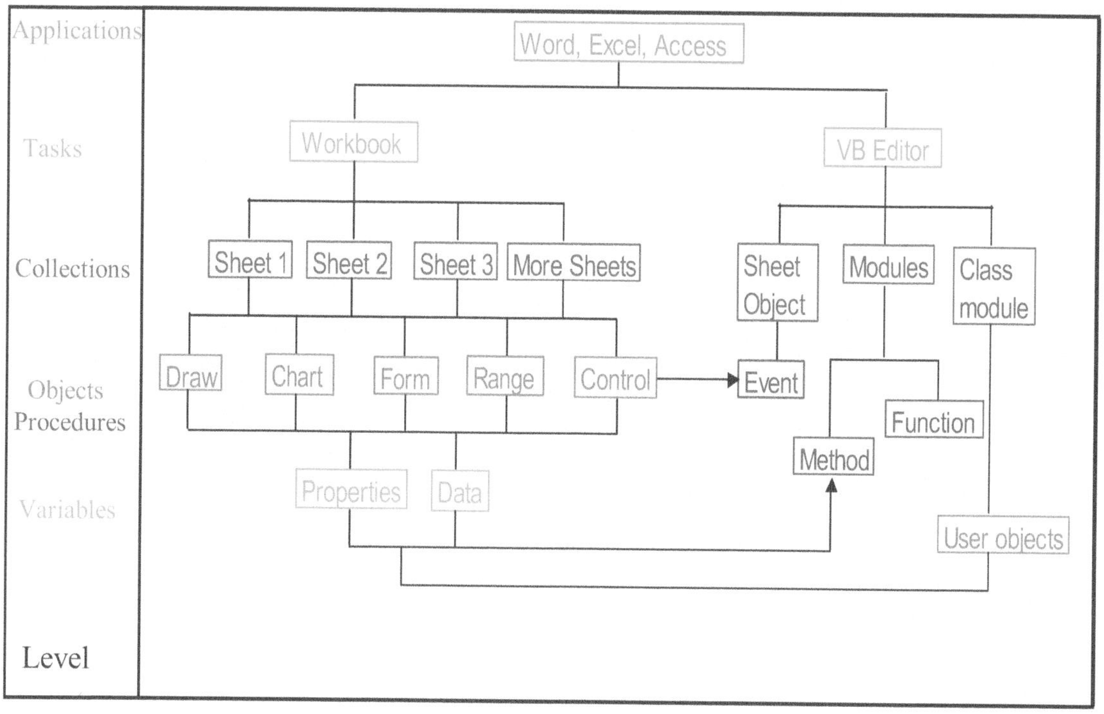

Figure 1.1.1: The "Visual Basic" object model

This first Visual Basic project looks at the organisation of objects (the object model, Figure 1.1.1), the toolbars or menus commonly used for designing VB projects (the environment), and the automatic writing of small programmes (recording macros).

The version of Visual Basic used herein (VB6 in VBA) is not wholly object oriented, although its successor (VB.NET) is. The latter is a complete redesign of the whole code, and conversion programmes are available. However, the object library is similar to all applications, and is structured as above.

All the codes for this book were originally designed for a PC environment that used a 1024 * 768 pixels monitor display setting with Normal size (96 DPI) fonts. The display is altered when different font sizes are used, as the default Excel worksheet cell size is changed. It is therefore necessary that the PC environment is set to Normal size fonts when using these projects.

A lower display setting than 1024 * 768 pixels will reduce the drawing area. It is then necessary to use the [Zoom] command in the [View] toolbar menu to increase the drawing area for some projects. The zoom reduction incurs a distortion of moving objects, and some projects may become unworkable for zoom reductions less than about 50%. A minimum display setting of 800 * 600 pixels with 60% [Zoom] reduction is however satisfactory.

1.1.2 The object model

Object oriented programs

Object oriented programming (OOP) has been in existence for several years. It is superceding the use of older procedural codes such as Fortran and Basic for the following reasons:

- It runs faster since there is efficient reuse of large portions of code (the objects).
- It enables users to interact with code by means of controls.
- It works with many applications such as Excel, Word, Access, and others. Note that other OOP such as Delphi may also communicate with MS applications.
- It enables the transfer of data between applications, platforms and systems.
- It is easier to write since it is developed in pieces (modules).
- It has an expandable object library for applications of increased size.
- It enables the immediate visualisation of pre and post processing of code.
- It enables small programmes to be recorded automatically.

The main reason why OOP has these advantages is because a major part of the code, the "Object Library" is already specifically prepared for each application. For example, the main 5 object collections used by Excel are drawings, charts, forms, ranges and controls. These collections are used for all the code examples used in this book, and no new object collection has been found necessary.

Perhaps the simplest explanation for OOP is that is similar to English grammar. The sentence:

I select a scatter chart

has a subject, verb, adjective and object. In OOP, the sentence becomes:

```
chart.scatter.select
```

and is controlled by the programme, a programme control event, or the User. Thus a grammatical object noun becomes an OOP "object", a verb becomes a "method" of the object, an adjective becomes a "property" of the object, and a subject noun becomes a "control" for the object. It is this simulation of natural language that eases the coding, and enables the object library to represent the whole application.

The VB object model shown in Figure 1.1.1 consists of several layers, beginning at the Application level. Although each of the levels may be considered an object, it is more appropriately described by its collective noun. A familiarity with applications is assumed, and the following discussion looks at the lower levels in succession. Note that there are several ways for issuing a command in Excel, and those detailed herein are based on the author's experience.

Tasks level

A Taskbar is the lowest line of any Windows display (although it may be moved elsewhere), as shown in Figure 1.1.2 below. It contains the [Start] icon at the left, and then the Task icons. The [Microsoft Excel – Project.xls] task refers to the Excel workbooks contained by the task. This however represents only part of the application, as the other major Task is the [Microsoft Visual Basic – Project.xls] or VB Editor (VBE). The latter is invoked by enabling the VB Toolbar and activating the [Visual Basic Editor] icon.

Using the workbooks alone will enable the drawing of objects, and the worksheet programming of worksheet cells. However, it is soon learned that worksheet programming is limited by reason of space in the formula bar, and the VBE provides far superior access for larger codes. The word "Project" is reserved for the description of a single project file and its associated code. Thus each of the code examples described in this book are called projects.

Note that multiple projects may be added to the application by use of so-called "Add-ins". They are accessed using the Excel commands:

[Tools, Add-ins] then checking the required programme

A particularly useful Excel "Add-in" is the data analysis programme, which calculates a variety of statistics. It is not however used in any of the present codes.

Collections level

Each workbook task contains a number of sheet objects. By default, the number of sheets is set to 3, but it may be altered using the Excel commands:

[Tools, Options, General, Sheets in new workbook]

(When using this command, it is always convenient to set the [Default file location] command towards future use.) Note that each worksheet has a name tab at the bottom left corner, and an appropriate worksheet name can be set.

The VBE has similar collections of windows. These are named as Sheet objects used for storing control event codes, Modules used for storing method and function codes, and Class modules used for storing user-defined object codes. As stated above, the latter are not used herein, but the former 2 collections are used throughout.

Objects level

All of the following projects in this book use the 5 object types of Drawings, Charts, Forms, Ranges and Controls in the worksheets. Although each object is associated with a specific toolbar ([View, Toolbars]), the [Insert, Chart] commands are used preferentially to the [Chart] toolbar.

The [Drawing] toolbar provides access to both 2 dimensional and 3 dimensional Drawing objects on the worksheet. The toolbar may be exercised by drawing the 2D and 3D examples given for this current project. The Chart commands provide access to 14 different Chart types, and up to 9 sub-types for each of these types. Only 3 different Chart types are used for these projects, corresponding to 1D, 2D and 3D data. They are exercised in the "Objects.xls" project. The Forms object type is not used extensively in these projects; forms are useful for controlled data input in commercial programming. There are however 2 commonly used "built-in" (intrinsic) forms in Excel, namely a Message Box and an Input Box. The former is demonstrated in this project, and the latter in the "Objects.xls" project.

The Ranges object describes the worksheet cells used for data. The object is edited using the [Formatting] toolbar, and its use is common throughout these projects. Finally, the Controls object types are invoked from the [Control] toolbox. Only 5 controls are commonly used in these codes, namely the Check box, the Command button, the Option button, the Spin button and the Scroll bar. Each control is associated with [Properties] and an [Event] code. They are readily accessed by entering VB Design mode (activating the [Design mode] icon) and right clicking the [Control] icon.

Procedures level

The procedures are the actual VB code used in the projects. For organisational purposes, the code in the control event procedures under the Sheet Objects is kept to a minimum. The code is used mainly to call method and function procedures developed in the Modules. It is also used to provide Message box warnings when irrelevant controls are operated.

Since Hydrology involves the intermediate calculation of many separate equations, it is useful to branch to a Functions procedure where all calculations are stored. The clarity of the Methods procedure is then maintained. The Functions procedure is labelled by the VB command Function Name(Variables) and the methods procedure by Sub Name(Variables). An example for separating subroutines and functions is given for the "Evap.xls" project. Note that a function procedure is not intended to interact with a worksheet, but only with the variables (the data) in the subroutine from which it is called.

1.1.3 The excel environment

The default Excel environment is sufficient for developing these projects. However, the addition of relevant toolbars facilitate the exercise. A knowledge of the coordinates of drawing objects on a worksheet is also required. These factors are now considered.

Toolbars

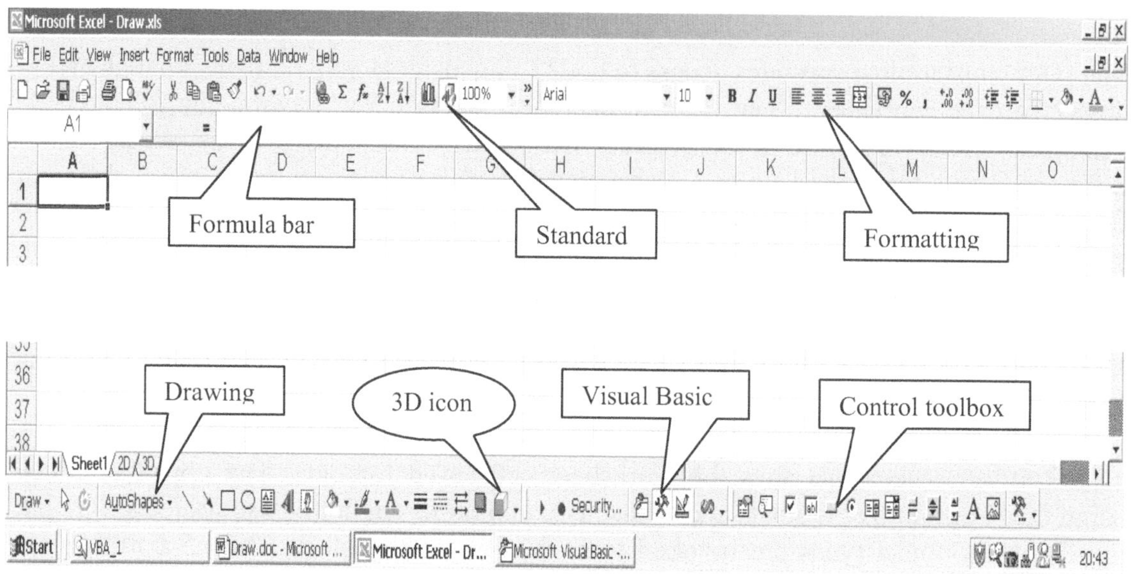

Figure 1.1.2: Toolbars commonly used for developing VB projects

The 5 supporting toolbars for VB development in Excel are shown in Figure 1.1.2. The toolbars are loaded using the Excel commands:

[View, Toolbars] then clicking the required toolbar

The chosen toolbars fit comfortably on the 3rd and bottom row of the Excel window. The formula bar is also included, as worksheet calculations are often needed to confirm parts of a program.

Display settings

The display settings are controlled by the operating system, and are accessed using the following [Windows 2000] commands:

[Start, Settings, Control Panel, Display, Settings, Advanced]

As noted above, a Screen Area of 1024 * 768 pixels (picture cells) is used for these codes, with Normal size fonts. With these settings, the [Column Width] and [Row Height] of a worksheet cell are automatically set in Excel to 8.43 (64 pixels) and 12.75 (17 pixels) respectively. (They are accessed by clicking on a row or column dividing line at the row or column alpha-numeric labels.) The former value of 8.43 is the average number of digits (from 0 to 9) of the standard font that will fit in a cell; it has nothing to do with any physical dimension in the worksheet. In contrast, the latter value of 12.75 represents the coordinate position of a graphical point at the first row in the vertical.

Excel sheet coordinates are referenced from the top left corner of the first worksheet cell (0,0). This first cell has a graphical coordinate width (abcissa) of 64/17 * 12.75 = 48 sheet units and a height (ordinate) of 12.75 sheet units. These units are important when it is desired to control the position of graphical objects on a worksheet, such as in the "Mano.xls" project. When the display settings are set to large fonts, the column width and row height become 8.11 (80 pixels) and 13.2 (22 pixels) respectively. A single cell's dimensions are then 80/22 * 13.2 = 48 sheet units and 13.2 sheet units. The vertical sheet unit therefore differs from that of the Normal font. A change in font therefore requires code changes for the positions of drawing objects.

Another important consideration for display settings is the number of rows and columns that are displayed on a worksheet. For a 1024 * 768 display setting, there are 1024/64 = 16 columns and 768/17 = 44.6 rows. For an 800 * 600 display setting, there are 800/64 = 12.5 columns and 600/17 = 35.3 rows. There is therefore a much larger drawing area for the higher resolution.

In summary, the adaptation of code from a normal to a large font requires changes in the positions of drawing objects. And the reduction of display settings requires the reduction of image size (by zooming) to show the same objects.

1.1.4 Recording a Macro

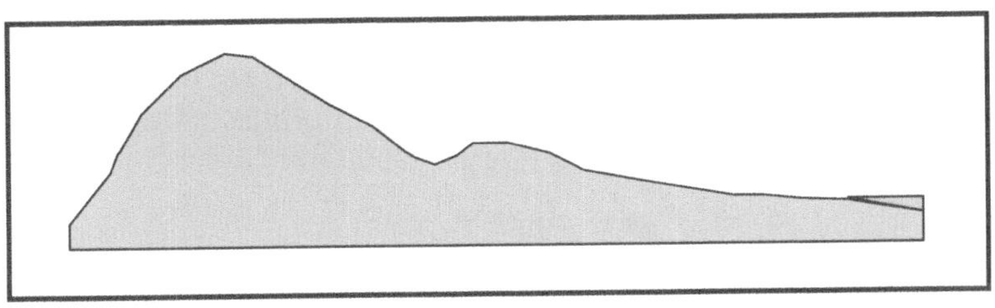

Figure 1.1.3: The 3D freeform

A simple way to begin coding in VB is to enable the automatic recording of key strokes and mouse clicks which are used in object creation and manipulation. The resulting code is called a [Macro] (an abbreviation for macroscopic, meaning a large scale as opposed to microscopic) and is written to the modules collection. The following example details the steps used in creating the 3D worksheet for this code. (The Excel commands are inserted in square brackets.)

1. Starting at the [Drawing] toolbar:

   [Autoshapes, Lines, Freeform]

The cursor should now be a cross, and the above shape (Figure 1.1.3) may be copied by clicking several similar points, and closing at the starting position with a double click.

2. Moving to the 3D icon on the [Drawings] toolbar, activate (click) the freeform then:

[3-D, 3-D settings, 3-D On, 3-D Color (Light Orange)]

3. One may now experiment by activating the [Tilt] icons to rotate the freeform. However, we wish to record these actions in a macro, so activate the [Record Macro] icon (the small, filled blue circle on the VB toolbar). A Message Box will appear to inform that [Macro 1] is being recorded in [This Workbook]. The macro will be written to [Module 1] in the VBE. Note that the small, filled blue circle changes to a small, filled blue square, and remains as such until the macro is terminated by clicking the square.

4. Activate the [Freeform], then tilt it several times to the left and right. Close and observe the macro by activating the [Visual Basic Editor] icon on the VB toolbar:

[Visual Basic Editor, Modules] then double click Module 1

The following VB code should be available:

```
ActiveSheet.Shapes("Freeform 1").Select
Selection.ShapeRange.ThreeD.IncrementRotationY -5#
Selection.ShapeRange.ThreeD.IncrementRotationY 5#
```

5. If we tilted the [Freeform 1] object to a maximum (180° of arc), we must repeat the IncrementRotation Y 5# command 36 times. Instead, we can write a small code to do this, which "loops" (repeats) the command 36 times. The macro may be copied and pasted, and the new code can be written in a new module called "Tilt" say, using the following commands in VBE:

[Insert, Module, View, Properties Window] then change the Name property to "Tilt"

The Do Loop is one of 2 very common procedures used in VB (the other being the If condition), and the new code becomes:

```
Sub HillTiltLeft()
        Sheets("3D").Shapes("Freeform 1").Select
        i = 1: Do Until i = 36: i = i + 1
        Selection.ShapeRange.ThreeD.IncrementRotationY -5#
        Loop
End Sub
```

6. This code may be Run by reducing the size of the VBE window, placing the cursor anywhere in the code and using the VB commands:

Run, Run Sub/User form

It is soon found that a "HillTiltRight" subroutine is required, and the rotation is too fast.

The rotation may be slowed by inserting code to wait for 1 second after each loop:

```
Application.Wait (Now + TimeValue("0:00:1"))
```

7. It would be more convenient to run this code from the worksheet itself. A control event is therefore required. A control object can be drawn on the worksheet by accessing the [Control] toolbox, activating the [Command Button] icon, and inserting the button to a relevant position. By right clicking the new control, the object [Properties] may be accessed, and the control [Caption] may be changed to say "TiltLeft". The [View code] command may also be accessed by right clicking, and it will open the VBE at the Sheet2 (3D) object. An event subroutine is already prepared and one simply inserts the macro name that is to be run, as follows:

```
Private Sub CommandButton1_Click()
        HillTiltRight
End Sub
```

8. There finally remains a form object to introduce on the worksheet as "Reset":

```
Private Sub CommandButton5_Click()
        Sheets("3D").Shapes("Freeform 1").Select
MsgBox "Click [3-D] on the Drawing toolbar then click [3-D settings]"
End Sub
```

1.2 OBJECTS AND PROCEDURES

1.2.1 Summary

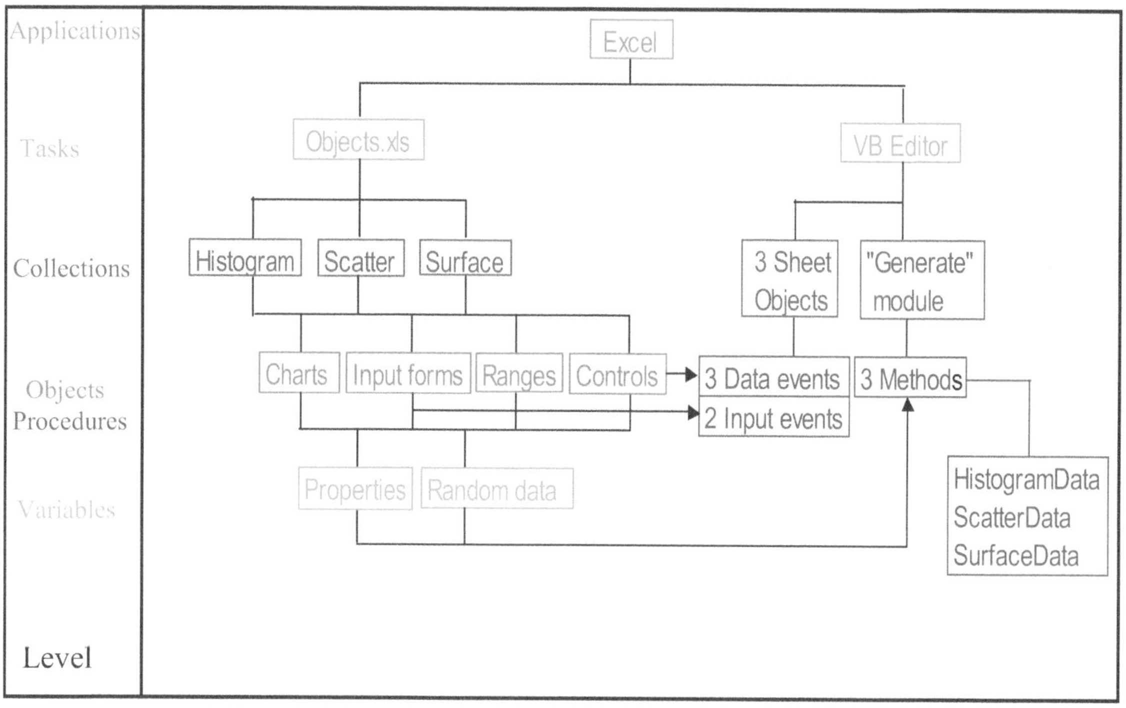

Figure 1.2.1: "Objects.xls" object model

This project demonstrates objects and procedures in more detail, since they are the main part of a programme. We begin with a more detailed description of the object library (Figure 1.2.1), then demonstrate sequentially 3 major chart object types used in engineering. The types are associated with the graphical display of 1, 2 and 3 dimensional data.

The data used are generated from random number selection using the VB intrinsic (built-in) Random function in the procedures. Since this method of data generation is used throughout the book, it is introduced at this early stage. A series of 1D data is generated in a procedure, sorted in ascending order, and plotted as a Column chart. The logical If condition is then used to sort the data into classes, and a histogram is plotted.

A series of data for 2D representation is generated, sorted, and plotted against a fixed, ascending abcissa on a Scatter chart. The slope and intercept of the sorted random data may be varied to simulate any particular distribution. This technique is used later in the Monte Carlo simulation of frequency distributions (Chapter 2, "EVA.xls").

Finally, a random array of data is generated and plotted as a 3D Surface chart and Surface plan chart. This illustrates the very useful ability of Excel to display surface data for say Digital Elevation Models.

1.2.2 The VBE object library

"Locals" Window and the Variables object

Type	*Size(bytes)*	*Values*
Boolean	2	True or false
Byte	1	0 to 255
Currency	8	About -10^{15} to 10^{15} with 4 decimal places
Date	8	Jan 1, 100 to 31 Dec, 9999
Double	8	About 10^{308} to 10^{-324} for positive and negative numbers
Integer	2	-32,768 to 32,768 whole numbers
Long	4	About -10^9 to 10^9 whole numbers
Object	4	Any type of object
Single	4	About 10^{38} to 10^{-45} for positive and negative numbers
String	Variable	Character data up to 2^{31} characters
Variant	22+	Default type for any data type (unavailable in VB.net)
Hyperlink	Variable	A file name on the same or networked system

Table 1.2.1: Types of variables

We begin this discussion by considering the object model from the bottom and moving upwards. A fundamental object is the variable representing data. As listed above in Table 1.2.1, a variable can take any character or numeric value. The storage size (in bytes) of a variable varies accordingly. It is therefore relevant to declare variables at the beginning of code to reduce the memory storage required, and speed the code operation. This is done by beginning a module with the statement Option Explicit, and following with declarations such as Dim x as Integer. If this is not done, the variable will default to a Variant type. (The latter is of course very useful when developing new code, however, the Variant variable type will become unavailable in VB.NET.)

The variable type can be viewed in VBE by accessing the Locals window with the commands View, Locals Window. The window automatically displays all of the declared variables in the current procedure and their values. For example, the following code may be typed into a new module:

```
Option Explicit
        Dim x As Integer, y As String, z As Single
Sub Test1()
        x = 500: y = "Dog": z = 25.6
End Sub
```

The code must then be run one line at a time by pressing "F8". It is seen, by activating the + sign in the Locals window, that the variable types are declared. Note that character strings are assigned using inverted commas, and the value may be edited (changed).

Locals Window and the Range object

Data variables are usually stored in worksheet cells, and are then called value properties of the Range object. For example, a data value in Cell A2 (where A is the column, 2 is the row) of the "Histogram" worksheet can be viewed in the Locals window using the following code. (Alternatively, the Range object may be set to a new object reference.)

```
Sub Test2()
        Dim Value As Single, NewObject As Object
        Value = Sheets("Histogram").Range("A2")
        Set NewObject = Sheets("Histogram").Range("A2")
End Sub
```

After running the above code in steps and activating the NewObject, all the Properties of the NewObject (i.e. the Range("A2") object) hierarchy are now visible. The complexity of the properties for the Range object model is now apparent.

Fortunately, Excel offers an easier method for accessing the commonly used properties and methods for a Range object. One simply right clicks on a worksheet cell or range of cells, and a menu appears which lists these properties and methods. Methods appear at the top of the menu [Cut, Copy, Paste, Paste Special, Insert, Delete, Clear], whilst Properties are contained in sub-menus [Format Cells, Pick From List, Hyperlink]. Note that the same applies for the Drawing, Chart and Control objects. (User form objects differ however, since they are accessed in the VBE.)

Remaining with the Range object in the VBE, it is now relevant to discuss 2 methods for accessing Ranges from code. As illustrated above, the first method is to define a range using its column then row name, as in Range("A2"). The other method is to use the Cells(2,1) object, which uses the row number variable before the column number. The advantage of the Cells object is that integer variables (as in Cells(i, j)) may be used in place of real integers, and the object may be used in Loops or If conditions. When the code requires a range of cells, either of the following may be used:

Range("A2:B31") or Range(Cells(2,1),Cells(31,2)

Note that the above range is otherwise known as an array or matrix. Excel code enables array manipulation such as multiplication, inversion, and transposition. As an example, a 9*9 array is generated in the "SurfaceData" code herein; it is assigned the single object name "Zarray".

In VB6, array variables may be named arbitrarily using the Option Base command. Thus a "Zarray(1 to 9, 1 to 9)" would have "Zarray(1,1)" as its first cell, using Option Base 1. (In VB.NET however, all arrays must have Option Base 0 for conformity with other codes.) Furthermore, an array size is limited to 5461 (74 * 74) elements in VB6 when it is assigned to a single name.

Object Browser

We have seen how the VBE Locals window is used to view an object's properties, and how the Excel object menu can expose commonly used properties and methods. For the sake of completeness however, the Object Browser is present in VBE to view all properties, methods and events both available to and used by Excel objects. The Browser is activated using the VB commands View, Object Browser, or clicking F2.

The first row of the Browser identifies the Library in which the object classes (collections) are stored. Thus <All Libraries> will show all available classes, whereas <VB Project> will show only the classes used in this "Object.xls" project. The second row of the Browser is used to search for information on a required class. For example, if the Range class is typed (with <All Libraries>) then activating the search (binoculars) will list all classes with the phrase "range" in it. These entries are located in the SearchResults section.

Members of the Range class can be located by activating the Range word in the Class column. And right clicking on a listed member, then clicking on Group Members will give an alphabetically ordered list of all properties, methods and available events for the Range class. Finally, selecting a member and pressing F1 will enable the VB Help programme that details the use of the member. Note that the Help programme is invaluable for developing code, since it gives code examples in addition to object explanations. It may also be accessed from within a code by positioning the cursor on the required word, and pressing F1.

Debug

Although the VBE has explicit Error commands for dealing with code errors, it is the author's experience that the Debug and Help programmes are adequate for successful projects. Both compilation and runtime errors are automatically trapped by VB, and explanations are given in the Debug programme. These explanations are usually sufficient to enable the rectification of code, or suggest the use of alternative code.

Perhaps the most useful VB command for debugging large codes is:

Debug, Run to Cursor

The cursor may be positioned at a relevant position in a procedure, and the code variables examined for correctness. This command provides a practical alternative to the Immediate or Watch windows.

1.2.3 "Histogram" worksheet

Code operation

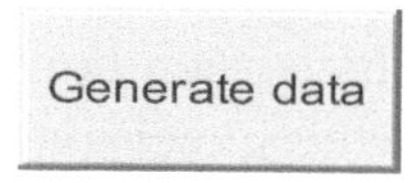

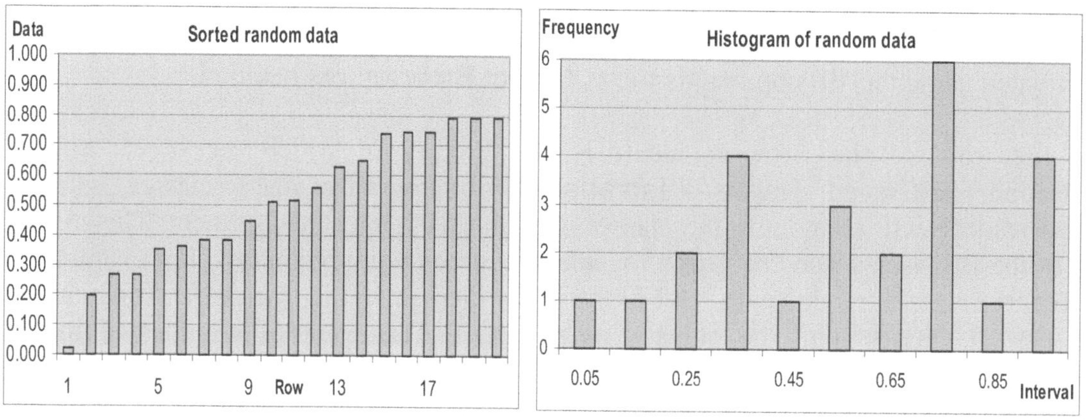

Figure 1.2.2: (a) Column chart of sorted, random data (b) Histogram of random data

The first example for plotting random data is that of a 1D Column chart (Figure 1.2.2). The data is generated and sorted in a "HistogramData" procedure in, say, a module named "Generate". It is loaded to the worksheet named "Histogram" using a single Command button whose caption is edited to be "Generate data". The Command Button has an associated event code that simply runs the procedure (as detailed in "Draw.xls").

To prepare this part of the project, the data must first be generated (as detailed below), then the chart objects may be drawn. Note that Row 1 of the worksheet is used for the data titles. (This may either be a coded entry, or a directly inserted worksheet entry as used for simplicity herein.) For Chart (a), the data in column A is first activated, then the Excel commands [Insert, Chart, Column] are activated. The Column chart object is drawn on the worksheet, and the abcissa (x axis) automatically takes sequential numerical values from unity. The Chart and Axes titles may be inserted by right clicking the chart, and using [Chart Options].

For Chart (b), the Frequency data in column C is activated, and another Column chart is inserted. Note that the abcissa is now 1 to 10 and requires editing to represent column B values. This is done by right clicking the chart, then using the Excel commands [Source Data, Series, Category(X) axis labels] then inserting = . At this point, one may activate the Column B (Interval) data and it will be automatically inserted to the abcissa after clicking OK. Finally, the titles may be inserted.

Note that a Chart has an additional method of [Order] to that of the Range object. The [Order] method enables the chart to be written as a layer on top of worksheet data ([Bring to Front]) or behind the worksheet data ([Send to Back]). In general, all graphical objects (drawings, charts, forms and controls) are layered and their order may be controlled singly (using the [Order] method), or in groups (using the [Group] method).

"HistogramData" subroutine

The procedure for this project can be prepared in 4 Steps, as follows:

1. Generate n random x values using the VB random function and looping.
2. Sort x ascending by recording a sort macro.
3. Generate intervals by writing a data series to a macro using the [Edit] command.
4. Calculate frequencies using a double If condition in a double loop.

For Step 1, we first use the VB Help programme to learn about random number generation. (Note that there are 2 separate Help programmes, one for Excel and one for VB). It is found that the following code will generate a random x value between 0 and 1 with 3 decimal places, and write it to the "Histogram" sheet:

```
Sheets("Histogram").Activate
        Randomize
        x = Int(((1001 * Rnd)) + 0) / 1000
Cells(i, 1).Value = x
```

There simply remains the insertion of a loop (as in "Draw.xls") to generate any number (n) of random values.

For Step 2, we must create a sort macro by activating Column A, and recording the Excel commands [Data, Sort, Sort Ascending, OK], as follows:

```
Columns("A:A").Select
Selection.Sort Key1:=Range("A2"),Order1:=xlAscending,Header:=xlGuess, _
     OrderCustom:=1, MatchCase:=False, Orientation:=xlTopToBottom
```

The recorded macro is much longer than necessary. It contains all the default properties of the method, as well as wasting runtime by selecting the object. We can edit the code:

```
Columns("A:A").Sort Key1:=Range("A2"), Header:=xlGuess
```

and it then runs faster. It is thus seen that although macros are convenient for identifying objects and methods, they use extra code and must be edited.

In Step 3, we again record a macro using the Excel commands [Edit, Fill, Series]. The new series requires a starting value that must be inserted and activated in the worksheet, a Step (interval) value (0.1) and a finishing value (0.95). Again, the recorded macro is cumbersome, and may be edited to the following:

```
Range("B2").Value = 0.05
Range("B2:B11").DataSeries Rowcol:=xlColumns, Step:=0.1
```

Finally, Step 4 will count the number of data that lie within a certain class interval. When learning to write code, it is helpful to write an uncoded statement for an operation first, thus:

Read the first x value and set a counter (q) to 0
If x is greater than or equal to zero and if x is less than 0.1 then q = q + 1

The double If condition is simply coded by repetition, and the code becomes:

```
j = 1: q = 0: Do Until j = n + 1: j = j + 1
        If Cells(j, 1) >= (p - 2) / 10 Then
        If Cells(j, 1) < (p - 1) / 10 Then
        q = q + 1
        End If: End If
Cells(p, 3) = q
Loop
```

There finally remains another loop to insert for the p interval values.

When the code is operated by the Command Button, it will be seen that it runs slowly due to the updating of data at each event. This updating can be suppressed by using the following VB command at the start of a procedure:

```
Application.ScreenUpdating = False
```

Alternative number of data (n)

The number of data points generated by the code may be changed by editing the n value in the procedure. However, activating the data series for each chart (by clicking on the blue columns) will show that Chart (a) has source data that is dependent on n. This source data must therefore be changed. By right clicking on the chart, the source data menu may be activated. The following entry:

[Values: = Histogram!$A$2:$A$26]

means that the data range A2 to A26 is being plotted from the Histogram worksheet. (The = sign denotes to Excel that the code is a formula, and the $ sign indicates that the cell numbers will not change if used as a formula elsewhere in the worksheet.) The [$A$26] value may now be changed to any alternative n value.

1.2.4 "Scatter" worksheet

Code operation

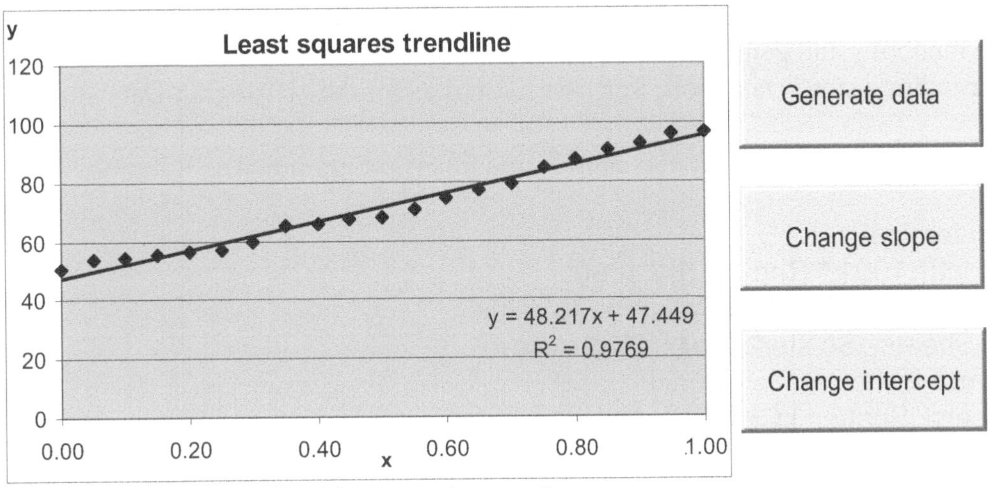

Figure 1.2.3: Scatter chart for a random series

The next example is that of a 2D Scatter chart (Figure 1.2.3). The data is generated and sorted in a "ScatterData" procedure in the "Generate" module. It is loaded to the worksheet named "Scatter" using a single Command Button whose caption is edited as "Generate data". The Command Button has an associated event code that simply runs the procedure.

There is only one set of random data generated, with values (F) from 0 to 1, and this is transformed to a linear distribution using the function "y = mF + c". The y data is then sorted in ascending order. The data, when plotted against fixed values of x (from 0 to 1), therefore becomes a random sample of a linear distribution given by "y = mx + c".

To generalise the distribution function, two Input boxes (intrinsic forms) are used for changing m and c respectively. The Input box properties may be accessed from the VB Help programme, and the entries may be included as an event procedure in the Sheet2 (Scatter) object, as follows:

```
Private Sub CommandButton2_Click()
        m = InputBox("Enter required slope", "Population slope", "50")
        Sheets("Scatter").Range("G19").Value = m
End Sub
```

Note that the m value is written to the "Scatter" worksheet, and is then available for the code. The Scatter chart is loaded after activating Columns A and B, and using the Excel commands [Insert, Chart, Scatter]. And a "Least squares fit" trendline is loaded by activating the chart data series, right clicking and using the commands [Add Trendline, Linear].

"ScatterData" subroutine

Only 3 Steps are required for this code:

1. Generate x values using an [Edit, Fill, Series] macro as for the "HistogramData".
2. Generate random y and generate the distribution function.
3. Sort y ascending using [Data, Sort, Sort Ascending] as for the "HistogramData".

The only new part of the code herein is Step 2, and this is programmed as follows:

```
Randomize
F = Int(((1001 * Rnd)) + 0) / 1000
m = Sheets("Scatter").Cells(19, 7).Value
c = Sheets("Scatter").Cells(19, 9).Value
y = m * F + c
Sheets("Scatter").Cells(i, 2).Value = y
```

Note how the variables m and c are assigned to the Input Box values via the "Scatter" worksheet. As for the "HistogramData" code, the above code must be looped for the relevant number of n values to be generated. And if alternative values for n are required, the Scatter chart source data must be edited.

1.2.5 "Surface" worksheet

Code operation

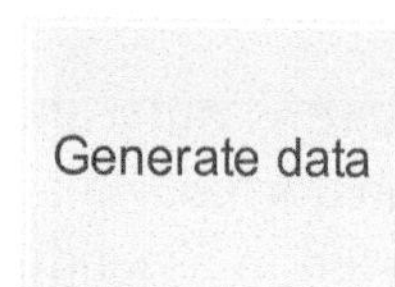

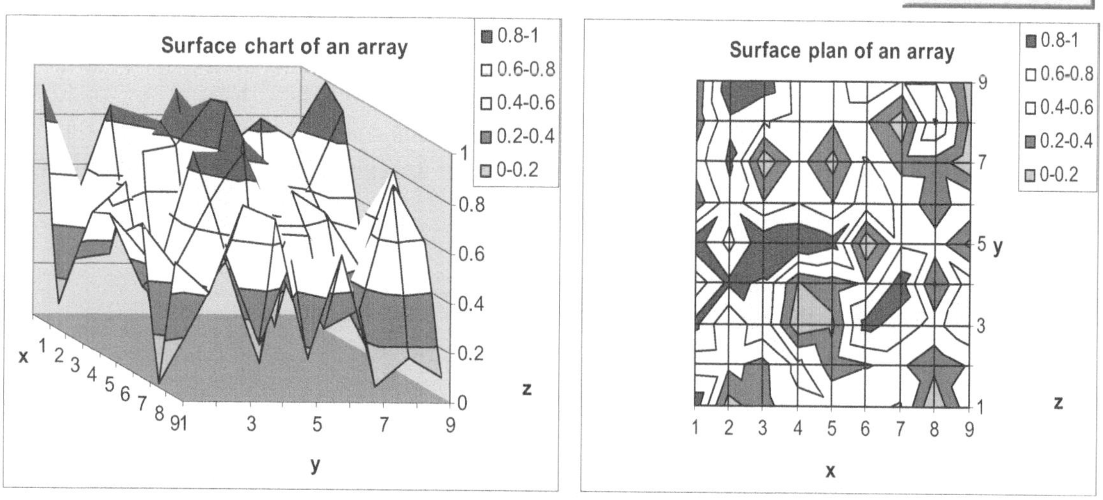

Figure 1.2.4: Surface chart for a random series

The final example is that of a 3D Surface chart (Figure 1.2.4). The random data is generated in a "SurfaceData" procedure in the "Generate" module. It is loaded to the "Surface" worksheet as a 9 * 9 element array, using a single Command Button. And the Command Button has the associated event code that runs the procedure.

The Surface chart is loaded after activating the array, and using the Excel commands [Insert, Chart, Surface]. Note that the Surface chart has 4 sub-charts, namely a 3D chart, the plan view for this chart, and corresponding wireframe models for these 2 charts. The example above shows the 3D and Plan view charts. Main and axes titles are entered as for the other chart objects, and the x and y axes values are set by right clicking on an axis value, and using the Excel commands [Format axis, Scale]. The z axis values are entered differently, by using for example the Excel chart commands [Source Data, Series, Name = "1"].

"SurfaceData" subroutine

Only 2 Steps are required for this code:

1. Generate random z and write to an array.
2. Write the array to the "Surface" worksheet.

The "Zarray" must first be declared using the dimension command. It can then be written using a double loop for the array components, as follows:

```
Dim Zarray(0 To 8, 0 To 8) As Single
n = 8
        j = -1: Do Until j = n: j = j + 1
        i = -1: Do Until i = n: i = i + 1
        Randomize
        z = Int(((1001 * Rnd)) + 0) / 1000
        Zarray(i, j) = z
Loop: Loop
```

"Zarray" is finally written to Cells B2 to J10 of the "Surface" worksheet. If alternative data is required, the n value must be changed and the Source data of the chart edited.

1.3 NUMERICAL OBJECTS – ESTIMATING THE PI CONSTANT

1.3.1 Summary

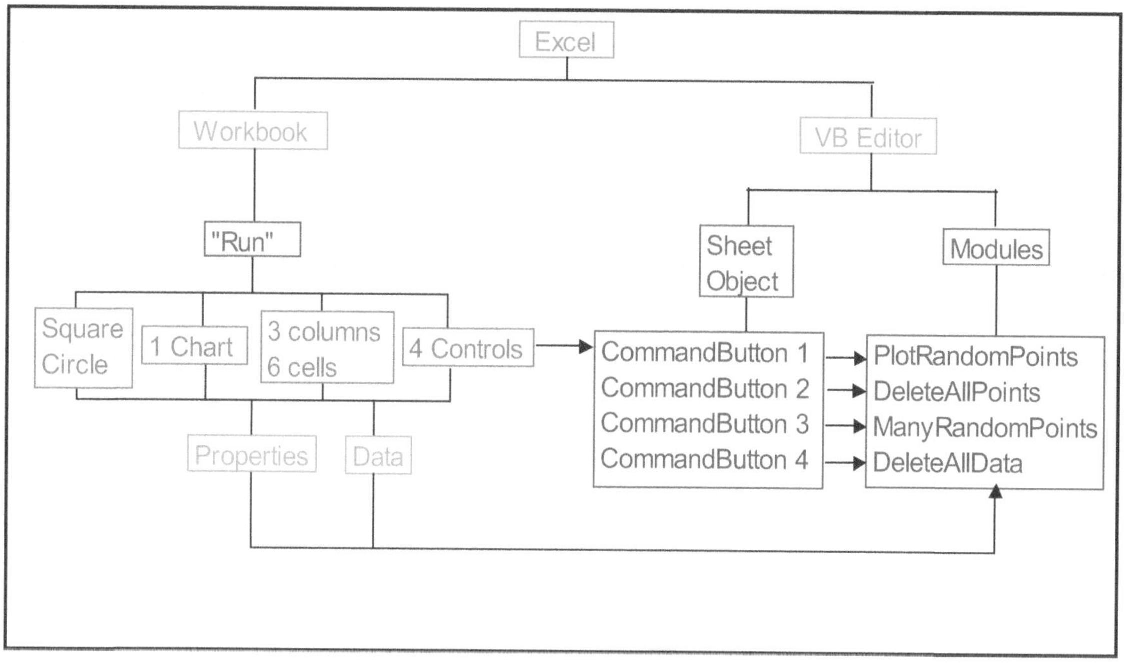

Figure 1.3.1: "PIMC.xls" object model

The main purpose for this project is to demonstrate how pairs of random numbers may be plotted as coordinates and used to numerically evaluate an area. The quadrant of a circular area within a square is used to evaluate the Pi constant, but any definite integral may be computed. The project further details the concept of the object model for a particular computation (Figure 1.3.1), and the important attributes of the Scope and Lifetime of objects are described.

The associated codes are used to amplify the various structures available in VB. Further examples for the Decision making structure (If condition) and Repetition structure (Loop condition) are given. And a new structure for abbreviating the reference to objects (With...End With structure) is introduced.

1.3.2 The scope and lifetime of objects

Variables

The Scope of a variable is its availability to the remainder of the VBE. For example, a variable that is created within a procedure is only available, and can only be changed within that procedure; it is called Local. A variable that is declared outside a procedure, in say the General Declarations section at the start of the module, is available to all procedures in the module; it is called Private and is the default scope which is also set by the Dim (Dimension) statement. A variable that is available to all modules must be declared using the Public statement instead of the Dim statement; it is now called Global. Note that if a module refers to functions in another module, the declared variables are Global to both modules.

The Lifetime of a variable refers to the longevity that a value is assigned. A variable exists for as long as a procedure has the control, or is running. When the procedure ends, the variable is released from memory and treated as having a zero value or zero length string. A variable can however retain its value while the project is open by declaring it as Static in place of the Dim statement used in the procedure. A static variable can be used, for example, to keep a running total of a value in a repeated procedure.

Note that if all variables in a procedure are required to be static, the Static keyword is inserted before the Subroutine declaration (as in Static Sub Test()).

Constants

Constants may be used instead of static variables, and they follow the same rules as for the Scope of variables. They are declared using the Const keyword, thus :

```
Public Const Pi as Single = 3.14159
```

After a constant is declared, it cannot be modified or assigned a new value.

Procedures

The same concept of Scope applies to a procedure. A Private procedure (as in Private Sub Test()) is only available to the module in which it is declared. Thus, when the Excel commands [Tools, Macro, Macros] are given, the macros in a Private subroutine are not listed. Note that the control event macros are always Private, as they refer exclusively to control objects.

A Public procedure (as in Public Sub Test()) is available to all modules. This is the default, and the procedures may be called from anywhere within the VBE without any reference to a module.

1.3.3 Code operation

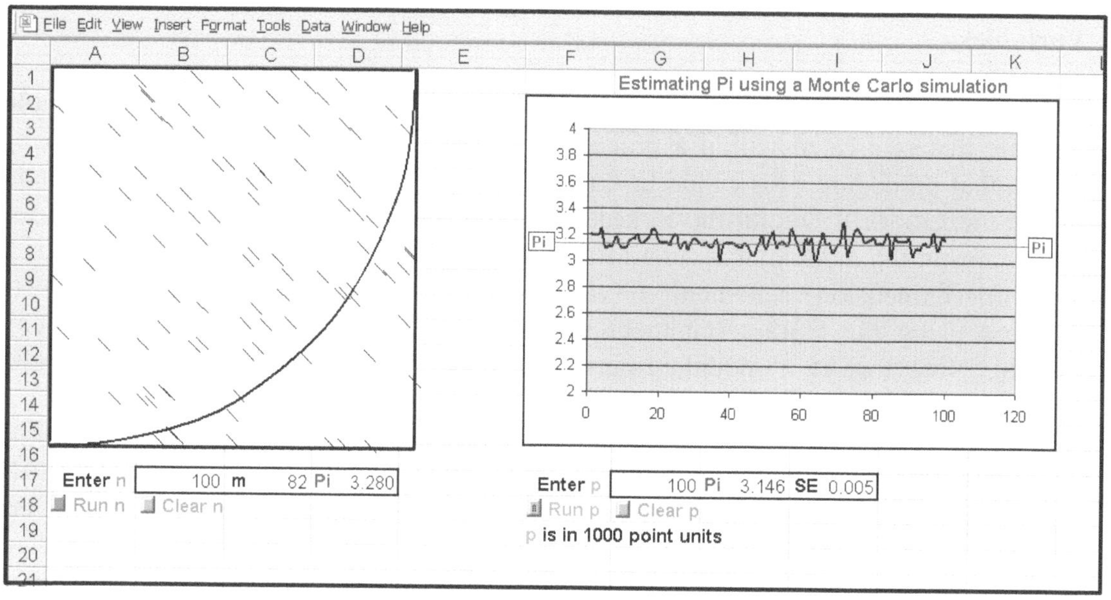

Figure 1.3.2: "Run" worksheet for the PIMC.xls code

The brief theory behind this code is that if n points are plotted at random on a square whose sides (D) are formed from the quadrant of a circle, then the chance (m) of n points falling within the quadrant is given by:

$$m/n = \text{area quadrant/area square} = (\pi D^2/4) / D^2 = \pi/4$$

Thus π may be evaluated by counting m (Figure 1.3.2).

The code is operated using the 4 Command buttons. On the left hand side, the "Run n" command is used to generate the n points entered in Cells(17,2). The points are plotted, and the number m is calculated and written to Cells(17,3). Pi is evaluated as m/n in Cells(17,4). The random points may be cleared using the "Clear n" Command Button.

It is soon found that even 1000 points give a poor accuracy for Pi, and a faster method is necessary. Since the runtime is slowed by drawing points on the worksheet, the code is rewritten without drawing. Values of p, where p is in 1000 point units, are entered using the "Run p" Command Button, and Pi with its standard error is estimated. This may be repeated without the need to clear the chart using the "Clear p" Command Button.

1.3.4 Subroutines

"DrawSquareQuadrant"

The code begins by drawing the square and enclosed quadrant as Freeform objects. Using the Excel commands [Autoshapes, Lines, Freeform], a macro can be recorded to draw any approximated rectangle in a new worksheet. The macro may then be edited to draw a suitably sized, accurate rectangle as follows:

```
With Worksheets("Run").Shapes.BuildFreeform(msoEditingCorner, 0, 0)
  .AddNodes msoSegmentLine, msoEditingAuto, 0, 200
  .AddNodes msoSegmentLine, msoEditingAuto, 200, 200
  .AddNodes msoSegmentLine, msoEditingAuto, 200, 0
  .AddNodes msoSegmentLine, msoEditingAuto, 0, 0
  .ConvertToShape
 End With
```

Note that the macro has produced a new code called a With...End With structure that obviates the need for repeatedly declaring the Worksheets("Run").Shapes object. The use of this structure is more efficient than using the Set variable = Object command given in the "Objects.xls" project.

The quadrant may be accurately drawn by positioning a divided circle ([Autoshapes, Flowchart, Flowchart: Or] over the square, and copying the arc using [Autoshapes, Line, Curve].

"PlotRandomPoints"

The random point plotting is done in 3 steps, as follows:

1. Check that n < 1000 using a one line If structure.
2. Generate x, y pairs of random data, count the data in the quadrant using a block If structure and plot the data.
3. Write data to the "Run" worksheet.

This procedure demonstrates the 2 forms of the If structure. The one line structure is used if the condition is a simple If...Then decision. Thus the code for Step 1 is:

```
If Range("B17").Value > 1000 Then MsgBox "Run p": End
```

This code will ensure that the slow data plotting will not occur if n > 1000, and the extra End command stops the code after displaying the Message box.

The Block If structure is used when alternative conditions are being tested (for which the Else If statement is used) or just an alternative condition is required. The following code operates in the x, y random data processing loop, and sets the counter j if the data point lies within the circle.

```
If (x ^ 2 + y ^ 2) ^ 0.5 <= 200 Then
j = 1
Else: j = 0
End If
```

Although almost any drawing object may be made small to simulate a point, the Line object is chosen for simplicity. The following code is derived from a macro to draw a line with origin at x and y.

```
With Worksheets("Run").Shapes.BuildFreeform(msoEditingCorner, x, y)
  .AddNodes msoSegmentLine, msoEditingAuto, x + 6, y + 6
  .ConvertToShape
End With
```

This code may then be used in the easy identification of lines for future deletion.

"DeleteAllPoints"

There are 2 main kinds of repetition structures, one of which is used with a counter (the Do Until… .Loop structure) and the other with an object (the For Each… .Next structure). Their differing purpose is best illustrated with code for deleting the random data points drawn in the previous subroutine. Firstly, a single event for deleting n points uses the following code:

```
Worksheets("Run").Activate
        i = 0: Do Until i = Range("B17").Value: i = i + 1
        Worksheets("Run").Shapes(17).Delete
Loop
```

This code requires some explanation for the Shapes collection. There are in fact 16 drawing shapes on the "Run" worksheet, before the new shapes of the data lines are drawn. These shapes may be identified using the following code:

```
Sub Test()
        Worksheets("Run").Shapes.SelectAll
End Sub
```

The new data lines are added as part of the Shapes collection, with the Index of 17 onwards. They are deleted in the same manner, hence the final shape to be deleted will become Shapes(17).

If the "Run n" control is now activated several times, then we no longer know how many shapes (lines) have been generated. The above code may be repeated several times, but it is easier to use the following code:

```
For Each Point In Worksheets("Run").Shapes
        If Point.Type = msoLine Then
        Point.Delete
        End If
Next
```

In this Loop, the "Point" variable (declared as an object) is identified as a Line, and all lines are deleted.

In summary, the Do Until....Loop may be used for a known number of repetitions, whilst the For Each ....Next Loop is used for a known object and unknown number of repetitions.

"ManyRandomPoints"

This subroutine may be written in 2 steps:

1. Generate pairs of random points using a double loop for 1000 points and for p.
2. Compute mean Pi and standard error (SE) using SE = $(\text{Variance} / p)^{0.5}$.

The code in Step 1 is similar to the "PlotRandomPoints" without the data plots, and the double loop begins with the statements:

```
p = 0: Do Until p = Cells(17, 7).Value: p = p + 1
i = 0: m = 0: Do Until i = 1000: i = i + 1: j = 0
```

Excel has more mathematical functions than VB, and these may be used in VB Code by applying the prefix Application.WorksheetFunction to the function name. Thus the average Pi value for the p samples is given by:

```
ybar = Application.WorksheetFunction.Average(Range("M:M"))
```

and the variance is calculated by:

```
Var = (Application.WorksheetFunction.Sum(Range("N:N"))) / (p - 1)
```

where column N has been coded to list the squared values of the residuals.

1.4 KINEMATIC OBJECTS – A MANOMETER SIMULATION

1.4.1 Summary

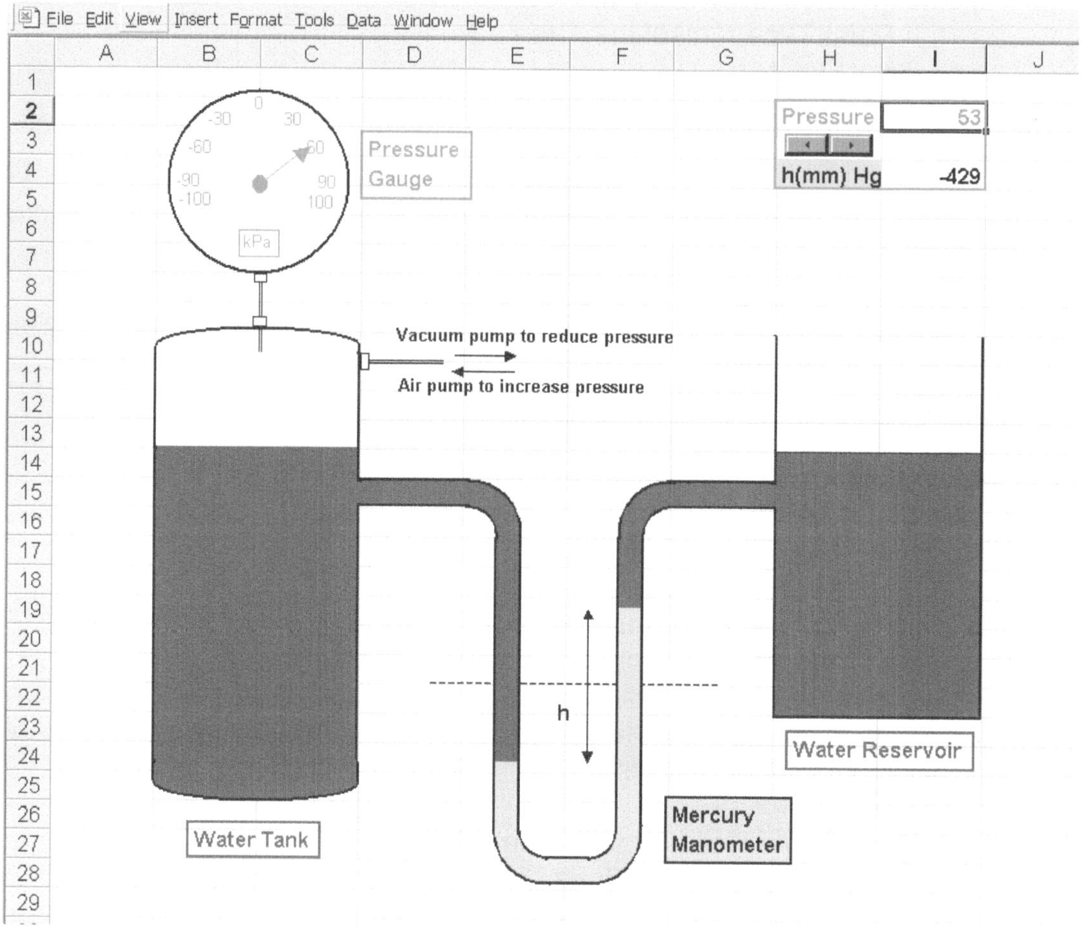

Figure 1.4.1: "Run" worksheet for the "Mano.xls" code

The purpose of this project is to practise drawing and manipulating objects, and to control them on the worksheet using VB code. The drawing skill requires a detailed knowledge of the Excel [Drawing] toolbar, and the manipulation of Autoshapes. Although curved objects may be freely drawn using Freeform lines, the manipulation of a curved shape is a faster and more presentable result (Figure 1.4.1).

The code to move an object on the worksheet requires the knowledge of object position, and this is demonstrated herein. The moving object is thus able to simulate real events, and can be used towards operating a "virtual laboratory".

1.4.2 Excel symbols used in manipulating drawing objects

Lines

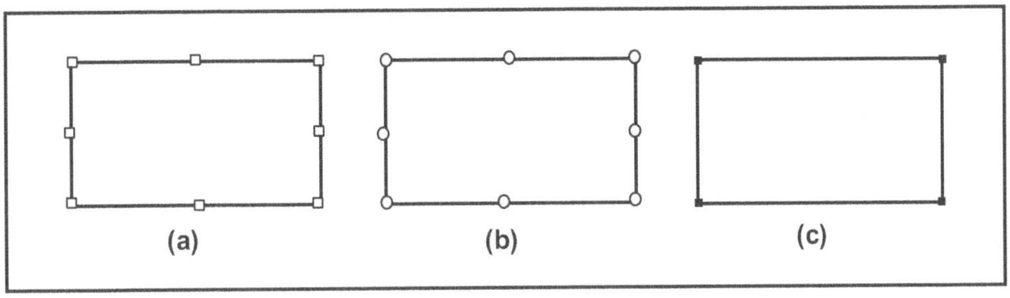

Figure 1.4.2: (a) [Selection] handles (b) [Rotation] handles (c) [Vertices] handles

The main part of the [Drawing] toolbar is the Autoshapes object. There are 2 kinds of autoshapes, namely those in the Lines sub-menu and those in the remaining sub-menus. The major difference between these groups is that the Lines objects can be edited at their vertices (called "nodes" in VB code).

When a Lines object is selected, its shape may be coarsely edited with the [Selection] handles (Figure 1.4.2(a)), or rotated about a [Rotation] handle by activating the [Free rotate] command on the [Drawing] toolbar (Figure 1.4.2(b)). At this stage, the object is still identified as a Line. However, if one right clicks on the object and selects the [Edit Points] method, the [Vertices] handles (Figure 1.4.2(c)) will appear. These are smaller handles than the selection and rotate handles, and allow the accurate editing of the object shape. Once edited, the Line becomes a Freeform object and is recognised in VB as such.

Shapes

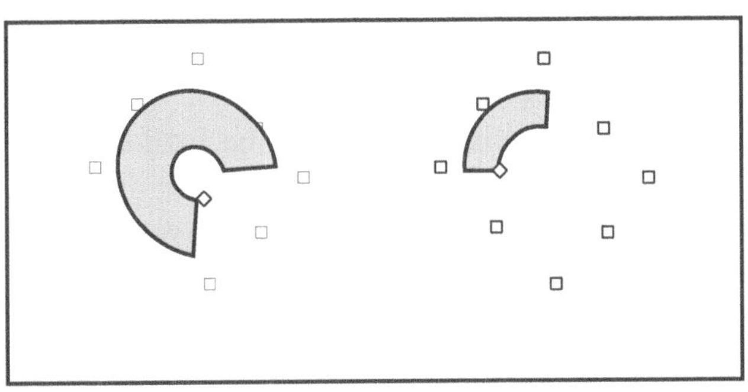

Figure 1.4.3: [Adjustment] handles

Many of the shapes (other than lines) have, in addition to selection handles, [Adjustment] handles (a small yellow diamond shape) that manipulate their appearance. For example, the Basic shapes. Block Arc shape in Figure 1.4.3 is used to draw the manometer tube elbows, and may be manipulated as shown. Note that a shape has the added property of a

[Border] line. This is useful in emphasising objects, and eliminates the need for the free drawing of borders.

Textboxes

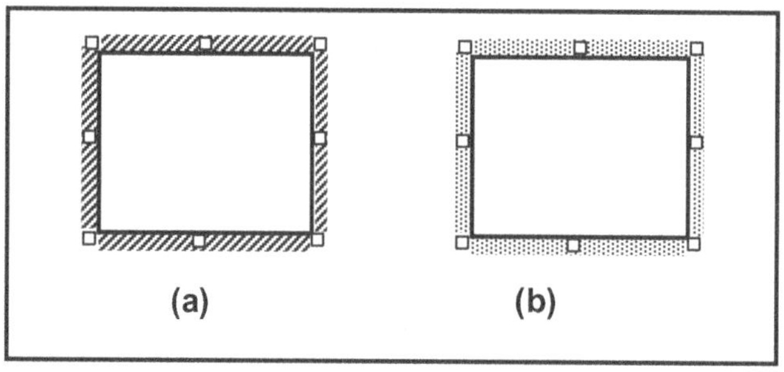

Figure 1.4.4: (a) Text-editing mode (b) Text box-formatting mode

The Text box is another frequently used graphical object in the [Drawing] toolbar. It has 2 editing modes. The text-editing mode (Figure 1.4.4(a)) is invoked when a Text box is first loaded, and enables text to be entered or changed inside the box. The Text box itself cannot be deleted whilst in this mode. The text box-formatting mode (Figure 1.4.4(b)) is invoked by clicking the box whilst in the text – editing mode. The border then changes to a dotted pattern, and right clicking the box enables the [Format Object] command.

1.4.3 Drawing the manometer apparatus

Initial design

There are 4 static, compound objects used to represent the apparatus of the "Run" worksheet, namely the Pressure gauge, Water tank, Water reservoir and Mercury manometer. And there are 4 objects that move following the activation of the single Spin Button, namely the Pressure gauge arrow, the left and right hand mercury column, and the mercury height difference (h) arrow.

By recording macros that change the dimensions of the Freeform objects (formerly the Lines object before editing the vertices) and the dimensions of the other Shape objects, it can be ascertained that VB code refers to the absolute (worksheet) positions of nodes for the Freeform, and relative positions for the Shape. For example, a recorded macro for positioning the top of the left hand mercury column ([Freeform 224]) looks like:

```
Sub Macro1()
  ActiveSheet.Shapes("Freeform 224").Select
  Selection.ShapeRange.Nodes.SetPosition 1, 213#, 273#
End Sub
```

Whereas a macro for reducing the length of a rectangle in the manometer ([Rectangle 205]) is given by:

```
Sub Macro2()
ActiveSheet.Shapes("Rectangle 205").Select
Selection.ShapeRange.ScaleWidth 0.81, msoFalse, msoScaleFromTopLeft
Selection.ShapeRange.ScaleHeight 0.89, msoFalse, msoScaleFromTopLeft
End Sub
```

Absolute positioning is usually necessary for moving objects, therefore the initial drawing design is based on using Freeforms for all moving objects.

1.4.4 Drawing the water tank and mercury manometer

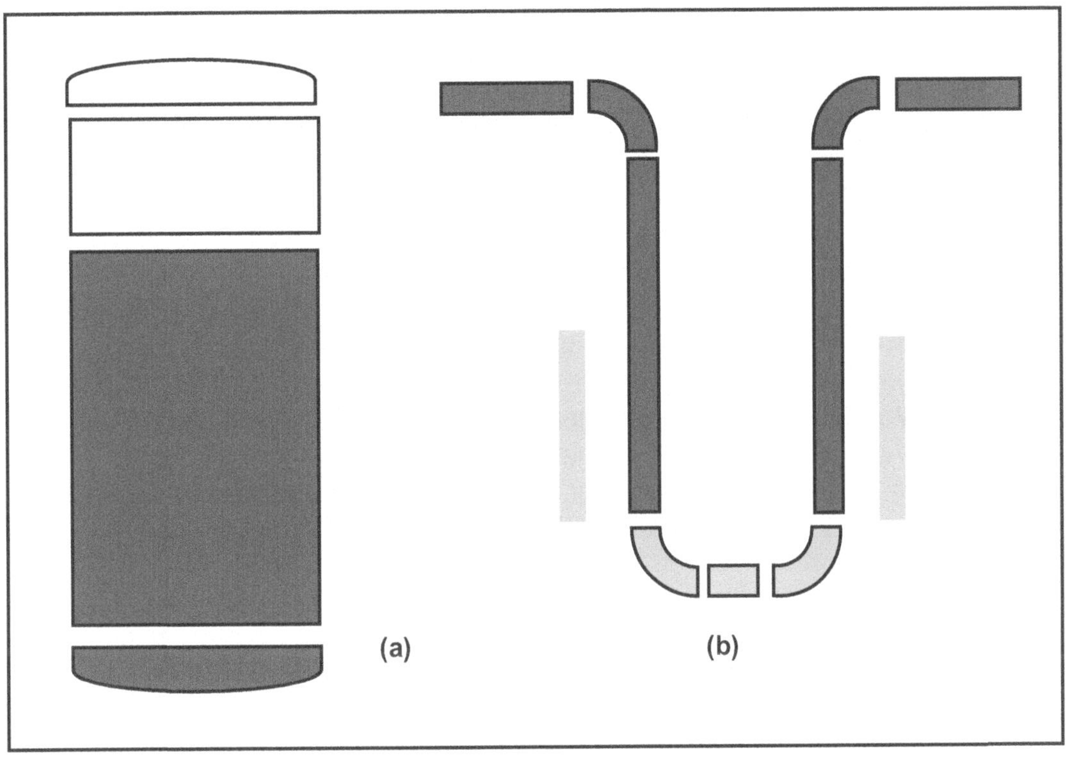

Figure 1.4.5: (a) Water reservoir (b) Mercury manometer

There are many ways to draw static objects. A general rule is however that when curves are involved, it is best to use the intrinsic shapes rather than using freeforms. For example, the Water tank and Mercury manometer are best built as illustrated in Figure 1.4.5.

A major advantage for using shapes is that [border] lines are available and less objects are used. Of course, when the shapes are joined, the double [border] line may not be desired. This is however easily removed by superimposing (using the [Order, Bring to the Front] method) a thick line of the desired colour and if necessary, editing at a [Zoom] magnification of 200%.

Note that the moving mercury column in Figure 1.4.5 is drawn using a thick line converted to a freeform, since the code only involves moving one node rather than the 2 nodes required by a freeform rectangle.

1.4.5 Subroutines

“ManometerGauge”

The first of 2 subroutines run by the Spin Button event code moves the Pressure gauge arrow as the Spin button is moved from +100 kPa to –100 kPa. Note that pressure entries may also be made in the usual way by editing Excel cells. This subroutine has 2 steps as follows:

1. Adjust the Spin Button reading to get correct negative readings.
2. Adjust Gauge direction arrow to align according to the Spin Button setting.

When the Spin Button is loaded to the worksheet, its [Scrolling] properties are set to a maximum value of 100 and minimum value of –100; the [LinkedCell] is then set to worksheet Range(“I2”). However, negative Spin Button values are represented by the number 65535 at –1 to 65436 at –100. The real negative value is therefore (Spin Button reading – 65536). The first step in the subroutine makes this adjustment as follows:

```
pos = Worksheets("Run").Range("I2")
neg = Worksheets("Run").Range("I2") - 65536
If pos <= 100 Then Worksheets("Run").Range("I2") = pos
If pos > 100 Then Worksheets("Run").Range("I2") = neg
```

The Pressure gauge dial displays the pressure (measured in kPa) in degrees of arc from zero to $\pm$ 100°. This corresponds to the pressure variation, therefore the latter is also a measure of degrees of arc for the gauge arrow, thus the variable “Degrees” is given by:

```
Degrees = Worksheets("Run").Range("I2")
```

Since VB calculates trigonometric ratios in radians, Step 2 of the code must convert the arrowhead coordinates to radians.

The arrow end position is set at the intersection of columns B and C, and rows 4 and 5. Its absolute position is therefore $x = 2 * 48 = 96$ sheet units, and $y = 4 * 12.75 = 51.0$ sheet units. Since the arrow length is chosen to be 30 sheet units, the arrowhead node is given by:

```
PosX = 96 + (30 * Sin(Radians))
PosY = 51.0 - (30 * Cos(Radians))
```

The position of the arrowhead can now be made by recording a short macro, and the code adjusted according to the above notes.

"ManometerLevel"

The remaining subroutine is in 3 steps, namely:

1. Adjust Manometer left (L) and right (R) mercury levels.
2. Adjust Manometer mercury level difference arrow.
3. Calculate h, the difference between the mercury levels.

In step 1, the distance moved by the level is a maximum of 5 rows for 100 kPa, therefore the incremental difference for 1 kPa is (5 * 12.75)/100 = 0.638. The left (PosL) and right (PosR) mercury levels are thus given by:

```
PosL = 277.2 + Degrees * 0.638
PosR = 277.2 - Degrees * 0.638
```

A macro may be recorded to change each level, and edited for the actual pressure reading.

The level difference arrow may be positioned in the same way. However, a problem exists as the arrow changes through the zero position. It can only exist for a minimum size of 4 Degrees. An If, Else If, Else condition must therefore be invoked as follows:

```
If Degrees > 0.4 Then
        Selection.ShapeRange.Line.Visible = msoTrue
ElseIf Degrees < -0.4 Then
        Selection.ShapeRange.Line.Visible = msoTrue
Else
        Selection.ShapeRange.Line.Visible = msoFalse: End if
```

Finally, the difference in mercury levels (h) is calculated from Figure 1.4.6 as follows:

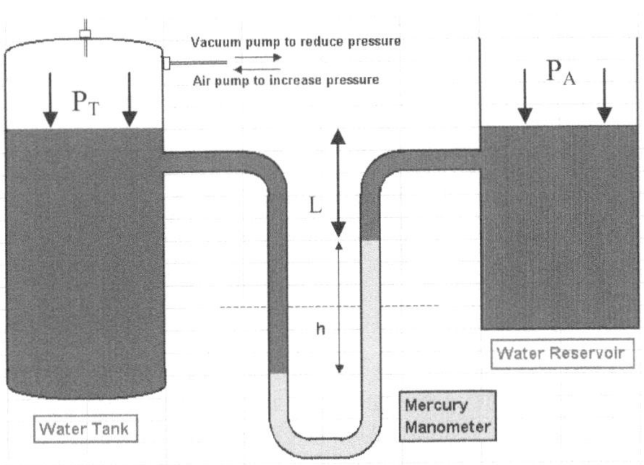

Figure 1.4.6: Pressure distribution

$$P_T + \rho_w g(L + h) = \rho_w gL + \rho_m gh$$

where P_T is gauge pressure and
$g = 9.81$ m/s^2
$\rho_w = 1000$ kg/m^3
$\rho_m = 13600$ kg/m^3

Thus h = -8.1 * kPa

1.5 FUNCTION OBJECTS - EVAPORATION AND POTENTIAL EVAPOTRANSPIRATION

1.5.1 Summary

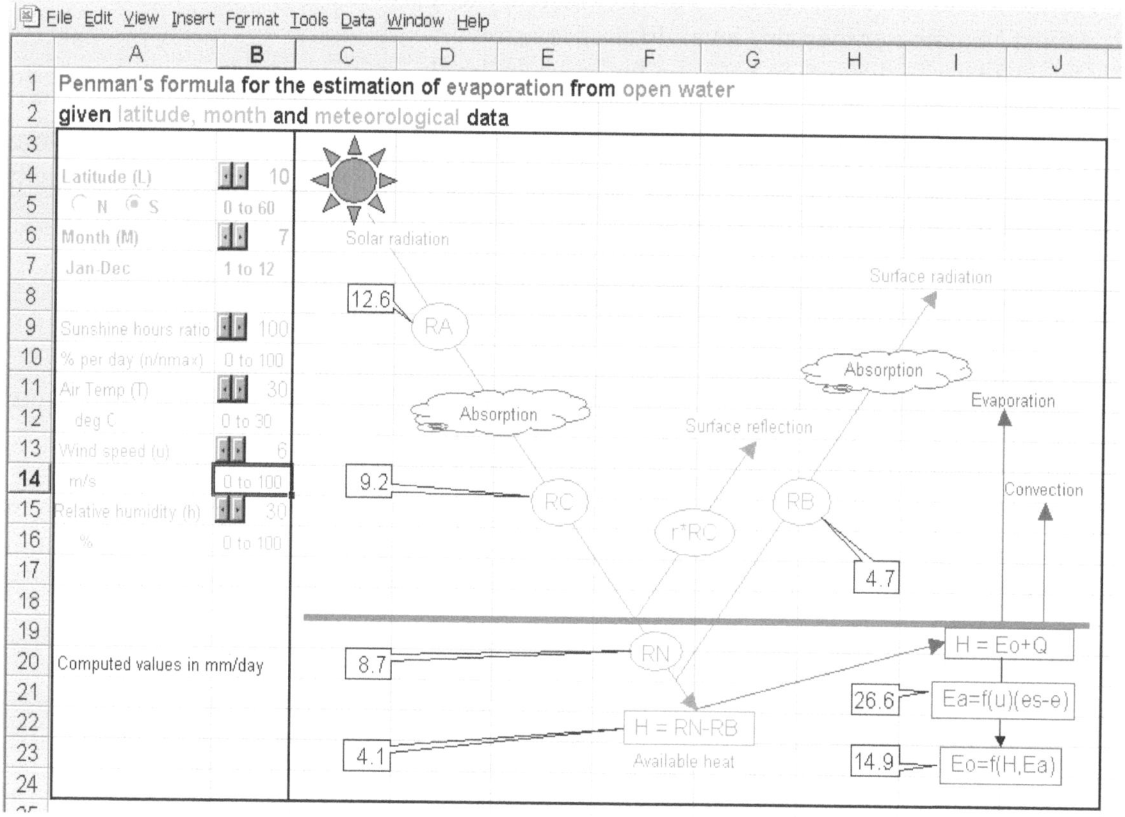

Figure 1.5.1: "Run" worksheet for the "Evap.xls" code

This code example is chosen to illustrate the use of the Function procedure in VBA. The computation uses Penman's formula (1948) for evaporation; this involves the calculation of at least 5 functions. The latter are best coded in a separate "Functions" module and then called by the "Penman" module.

Penman derived a classical equation for estimating evaporation from any global, open water surface using standard meteorological observations. The first part of the theory uses a Conservation of Energy approach towards estimating the resultant Solar radiation energy (H) at a water surface. The second part of the theory considers that H is dissipated in both evaporation (E_o) and convection (Q) at the water surface. By assuming that the latter two processes are governed by the same physical mechanism, a final formula for E_o is reached (Figure 1.5.1).

The formula has been adapted for land surfaces in which evaporation combines with the transpiration ("breathing") of plants. For unlimited soil moisture, the total moisture dissipation is called "potential evapotranspiration".

1.5.2 Code operation

There is only one subroutine named "LocalEvaporation" that operates this code, and this is located in a module called "Penman". The subroutine is called by all of the 6 data Spin buttons and the 2 "Latitude" Option buttons. Thus, when the data is changed by any Control button, the code associated with the button activation (located in the Sheet1 (Run) object) runs the subroutine "LocalEvaporation". This is simply achieved by writing the following code using SpinButton1 as an example:

```
Private Sub SpinButton1_Change()
LocalEvaporation
End Sub
```

The Spin Button properties must be set. They are accessed by the following commands:

[VB Design mode. (Right click on Spin button.) Properties.Categorized]

The important properties to set are [Max, Min, SmallChange]; these values are determined by the respective data. The properties of [LinkedCell, BackColor] are also set, where [LinkedCell] is the cell that displays the button value and [BackColor] is the button's colour.

In addition to using the Control buttons to change data, the normal method of Excel data entry for cells may be used. Thus decimal entries for data may be applied. Note that this direct method of data entry causes the Spin Button value property to change, and therefore also runs the code.

The data are entered in the control panel at the left of the "Run" worksheet, and consist of the 6 independent variables:

$$E_o = f\,(L, M, n/nmax, t, u, rh)$$

In some cases, the data may be measured in alternative terms. For example, there may be a direct measurement of average, daily sunshine hours (n). In this case, nmax must be computed using the "nmax function", as described later. Similarly, the average vapour pressure of air (VP) may be measured instead of the relative humidity (rh). The latter is then calculated as rh = VP/SVP = e/e_s, where e_s is computed from the "SVP" function. Finally, the daily average wind speed may be estimated for a non-standard height above sea level. An empirical conversion between a wind speed u_o (at height z_o) and wind speed u (at height z) is given by Wilson (1990) as:

$$u/u_o = (z/z_o)^{0.15}$$

where $z > z_o$.

1.5.3 Theory

Conservation of Energy

There are 5 stages used in estimating the available radiated heat energy (H) at the open water surface from the boundary solar radiation, as follows:

1. Boundary solar radiation (R_A) is the short wave solar radiation flux at the outer limit of the atmosphere. This quantity is a function of global position and time, and has been tabulated in several publications. For convenience, the tabulation given by McCulloch (1965) for Latitude (L) and Month (M) gives the flux directly in terms of mmH_2O/day and is reduced in this code to a 3 dimensional cubic trend equation.

2. Energy loss due to cloudiness is estimated using the ratio n/n_{max}, where n is the observed number of daily sunlight hours for an observation period (a month say), and n_{max} (= f(L,M)) is the tabulated maximum possible average hours of daily sunshine (MAFF, 1967). The solar energy which is reduced by the cloudiness (R_C) is then given empirically by:

$$R_C = R_A\,(0.18 + 0.55\; n/n_{max})$$

The nmax tabulations are again reduced to a 3 dimensional cubic trend equation for this code. Note that if n/nmax is unavailable, it may be estimated (Wilson, 1990) using:

$$n/n_{max} = (1 - \text{fraction of daily cloud cover})$$

3. Energy loss due to reflection at the water surface is estimated using a reflection coefficient (r) of about 0.05 (otherwise known as the "albedo"). The solar energy which is reduced by the albedo (R_N) is then given empirically by:

$$R_N = R_C\,(1 - r)$$

4. Energy loss due to long wave radiation of the water surface is estimated using physical laws which assume the water body is a "black body radiator" and radiates according to its absolute temperature (273 + t°C), thus:

$$\text{BBRadiation}(t) = ((273 + t) * 0.01) \wedge 4 * 0.195$$

The BBRadiation is reduced by the ambient vapour pressure (which is equal to the product of the relative humidity (rh) and saturated vapour pressure (SVP)), and the cloudiness (n/n_{max}). A final empirical estimate (R_B) is then given by:

$$R_B = \text{BBRadiation}(t) * (0.56 - 0.09 * (rh * SVP(t)) \wedge 0.5) * (0.1 + 0.9 * n/nmax)$$

5. Available heat energy is finally estimated in units of mmH_2O/day as:

$$H = R_N - R_B$$

In summary, the functional expression for H is given by:

$$H = f(L, M, r, n/nmax, t, rh)$$

1.5.4 Dissipation of Energy

This available heat energy (H) is dissipated as evaporation (E_o) and convection (K), thus:

$$H = E_o + K \quad \text{or} \quad H/E_o = 1 + K/E_o$$

Two assumptions are made to evaluate this expression. The first is that the transport of vapour (E_o) and the transport of heat (K) are governed by the same physical processes (namely boundary layer turbulence). An empirical representation for E_o has been derived as:

$$E_o = f(u)\,(e'_s - e)$$

where u is the wind velocity measured at a standard height of say 2 m above the water surface (it is zero at the surface), e'_s is the SVP of the warmer air above the surface, and e is the actual vapour pressure of this warmer air (e/e'_s is the relative humidity). It follows by the first assumption that K can be given by:

$$K = f(u)\,(t'_s - t)$$

where t represents the respective temperatures. Thus:

$$K/E_o = f(u)\,(t'_s - t) / f(u)\,(e'_s - e) = \gamma\,(t'_s - t) / (e'_s - e)$$

whence $$E_o = H / [1 + \gamma\,(t'_s - t) / (e'_s - e)]$$

where γ is known as the psychrometric constant (0.485 mmHg/°C) and results from the fact that vapour pressure is directly proportional to temperature.

The second assumption is now introduced by replacing t_s with t, since they are close in value (t represents the temperature of warmer air in the boundary layer above the water surface, which is of temperature t_s). Referring to Figure 1.5.2, the gradient de/dt (Δ say) near t_s is given by:

$$\Delta = (e'_s - e_s) / (t'_s - t_s) \approx (e'_s - e_s) / (t'_s - t)$$

Thus:

$$E_O = H / [1 + (\gamma / \Delta)\,(e'_s - e_s) / (e'_s - e)] = H / [1 + (\gamma / \Delta)\,((e'_s - e) - (e_s - e)) / (e'_s - e)]$$
$$= H / [1 + (\gamma / \Delta)\,(1 - (e_s - e) / (e'_s - e)]$$

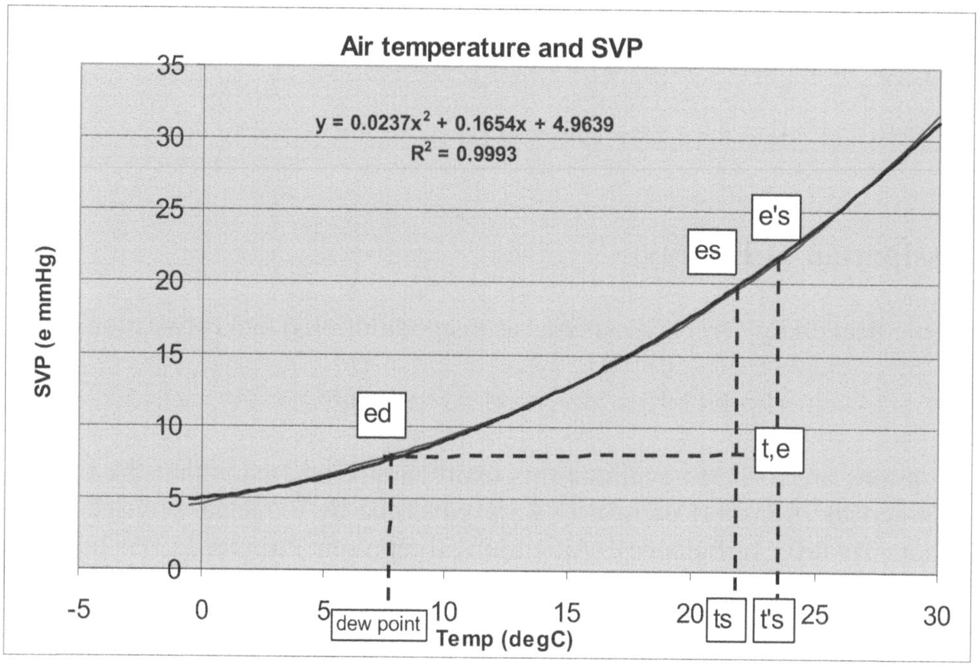

Figure 1.5.2: Saturation vapour pressure against temperature

The evaporation energy for the hypothetical case of equal temperature of air and water (E_a say) is given, using the above empiricism, as:

$$E_a = f(u)\,(e_s - e)$$

Thus: $$E_o = H / [1 + (\gamma / \Delta)\,(1 - E_a / E_o]$$

whence: $$E_o = (\Delta H + \gamma E_a) / (\Delta + \gamma)$$

In functional form, $$E_o = f\,(H, u, e/e_s, e, \Delta)\ \text{mmH}_2\text{O/day}$$

The evaporation energy, E_a is evaluated by solving for $e_s(t)$ from a SVP versus T curve (Figure 1.5.2, MAFF, 1967) and using this with measured relative humidity to solve for e. The empirical equation used for E_a (Penman, 1948) is:

$$E_a = 0.35\,(0.5 + 0.54u)\,(e_s - e)$$

where u is measured in m/s at 2 m above sea level, and e, e_s is in mmHg. The gradient Δ is obtained by differentiating the SVP versus t curve.

1.5.5 Potential evapotranspiration

The above considerations refer to evaporation from an open water body. Evaporation from a natural land surface, such as grassland, must also include the effects of transpiration and soil moisture availability. The physical effect of transpiration involves both the vegetation water uptake and exposure of water vapour at the vegetation surface for evaporation. Inevitably, there will be a loss of moisture by vegetation consumption and growth. Since both evaporation and transpiration are a dependent process, they are collectively referred to as evapotranspiration. A first approximation towards computation is made by assuming an unlimited soil moisture supply. Such a computation is then referred to as "Potential evapotranspiration" (PE).

Many formulae have been proposed for PE (Shaw, 1994 and Wilson, 1990). A method developed by extending the Penman formula has however been proposed for use in the UK (Meteorological Office, 1981). The conservation of energy is best presented by comparison with the open water case, where H is given by:

$$H = R_N - R_B = R_C\,(1\text{-}r) - R_B = 0.95R_C - R_B$$

For grassland say, r is now given as 0.25 and R_B is 0.95 that for open water, thus:

$$H_{PE} = R_N - R_B = R_C\,(1\text{-}r) - R_B = 0.75R_C - 0.95R_B$$

The dissipation of energy function for the open water case is given by:

$$E_a = 0.35\,(0.5 + 0.54u)\,(e_s - e)$$

For grassland say, the extra roughness at the boundary layer is included in f(u) as follows:

$$E_{aPE} = 0.35\,(1 + 0.54u)\,(e_s - e)$$

Thus: $$E_{oPE} = (\Delta H_{PE} + \gamma E_{aPE}\,)\,/\,(\Delta + \gamma)$$

The above method has been further extended to include the influence of available soil moisture. The computations then become locally intensive on the geography and geology of a catchment. An example of such computations for the UK is given by the "Meteorological Office rainfall and evaporation calculating system" (Wales-Smith, 1975), and is being continually updated.

1.5.6 Modules

"Penman"

The object of this code is to demonstrate how a Method procedure is simplified by using Function procedures. We begin by writing the main subroutine in the "Penman" module called "LocalEvaporation". The code begins by activating the "Run" worksheet and proceeds in 3 steps as follows:

1. Declare variables.
2. Calculate evaporation.
3. Write data to the "Run" worksheet.

In Step 1, the 6 data variables are declared from the "Run" worksheet using, for example:

```
L = Cells(4, 2).Value
```

The variables Gamma and r are also declared (Gamma = 0.485, r = 0.05).

In step 2, the 7 steps for calculating RA, RC, RC, RB, H, Ea and Eo are defined as described in the "Theory" section. For example, we have:

```
1: RA = SolarRadiation(L, M)
```

where SolarRadiation(L,M) introduces the SolarRadiation function with its associated variables L and M (described below). The next line of code is:

```
2: RC = RA * (0.18 + 0.55 * n_nmax)
```

In step 3, the 7 computed variables are written back to the Callout AutoShapes in the "Run" worksheet. For example:

```
Cells(8, 3).Value = RA
```

Finally, all variables are declared at the start of the module to reduce the memory required for the code.

"Functions"

This module contains 5 formulae derived from relevant data published in the literature. Since the functions involve fitting cubic polynomials and cubic trend equations, they are directly quoted below without the associated detail.

```
Function SVP(t) As Single
SVP = 0.0237 * t ^ 2 + 0.1654 * t + 4.9639
End Function

Function Delta(t)
Delta = 0.0119 * t + 0.1654
End Function

Function SolarRadiation(L, M) As Single
If Sheets("Run").OptionButton2.Value = True Then M = M + 6
If M > 12 Then M = M - 12
SolarRadiation = 13.427 + 0.8867 *M - 0.2299 * L - 0.17 * M^2 - 0.002593 * L^ 2
+ 0.1082 * M * L + 0.0088 * M ^ 3 + 0.00001067 * L ^ 3 - 0.008652 * M ^ 2 * L +
0.00002112 * M * L ^ 2
If SolarRadiation < 1 Then SolarRadiation = 1
End Function

Function nmax(L, M)
If Sheets("Run").OptionButton2.Value = True Then M = M + 6
If M > 12 Then M = M - 12
nmax = 12.397 + 0.05348 * M - 0.1721 * L - 0.05616 * M ^ 2 - 0.00006476 * L ^ 2
+ 0.07917 * M * L + 0.004527 * M ^ 3 + 0.000003244 * L ^ 3 - 0.006108 * M ^ 2
* L - 0.00003786 * M * L ^ 2
End Function

Function BBRadiation(t)
BBRadiation = ((273 + t) * 0.01) ^ 4 * 0.195
' MsgBox "BBRadiation"
End Function
```

The "SVP", "SolarRadiation" and "nmax" functions are not exact; they fit the data with a "goodness of fit" coefficient (R^2) of 0.99, 0.95 and 0.92 respectively. Furthermore, the "SolarRadiation" function must be corrected for trended negative values, using an If condition. Note that a southern latitude is computed by phase shifting the northern latitude functions by 6 months.

Finally, the computed value of a function may be readily observed by using a Message Box command as follows:

```
Sub TestSVP()
x = SVP(20)
MsgBox x
End Sub
```

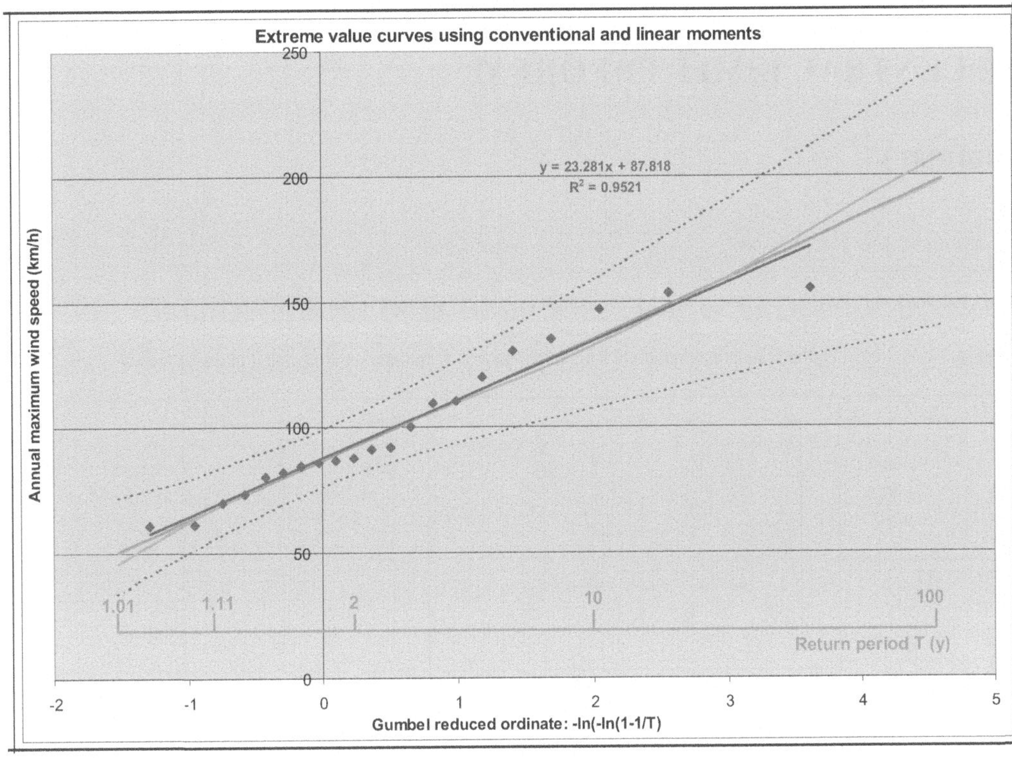

2 Statistics and Geostatistics

2.1 THE CENTRAL LIMIT THEOREM

2.1.1 Summary

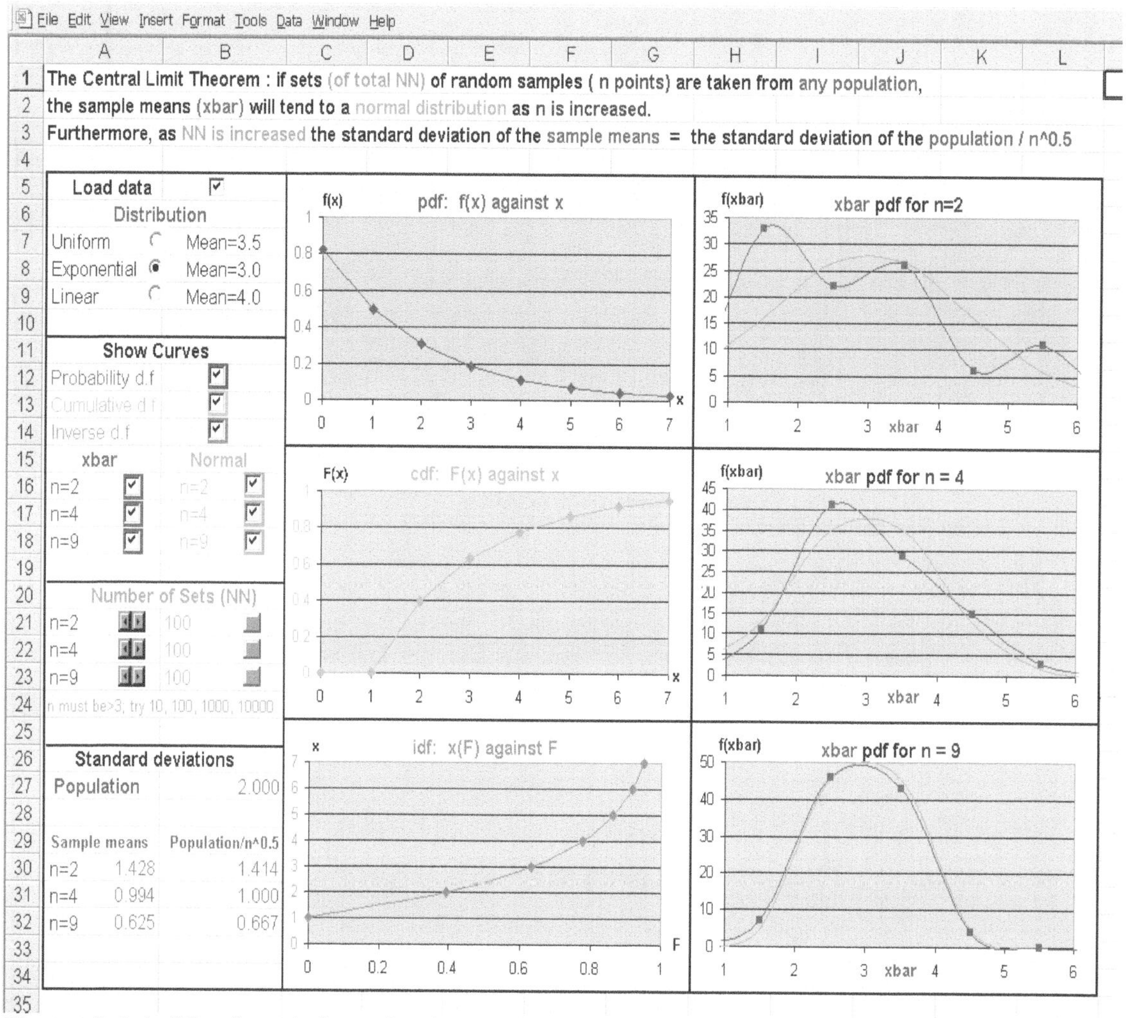

Figure 2.1.1: “Run” worksheet for the “CLT.xls” code

The Central Limit Theorem (CLT) is one of the most fundamental results in probabilty and statistics. Formally, it is stated that if sets (of total NN say) of random samples (X_1, X_2, ... X_n) are taken from any population (with mean μ and finite variance σ^2), then the probability distribution function (pdf) of $\overline{X}$ (the sample mean, pronounced xbar), approaches a normal distribution as $n \to \infty$, thus

$$\text{Limit}_{\,n \to \infty}\ \text{pdf}\,[\overline{X}\,] = N(\mu, \sigma^2/n)$$

This present code demonstrates the CLT for 3 different pdf's (Figure 2.1.1). It also demonstrates that as the number of samples (NN) is increased, the standard deviation of $\overline{X}$ approaches the standard deviation of the population divided by $n^{0.5}$.

2.1.2 Code operation

The code has 4 areas in the control panel. The first area loads the data and chooses one of 3 distributions. The second area identifies the curves used in the charts, by switching them on or off. The third area enables a user to select the number of data sets (NN) used for analysis. And the fourth area provides the population and sample statistics. The associated control subroutines are written in the Sheet1 (Run) object, and are now described.

When the "LoadData" Check box is clicked on, the "Uniform" Option button is turned on by the Check box control code, and this in turn operates all the subroutines necessary to generate the full set of "Uniform" curves. When the "LoadData" Check box is clicked off the chart source data is reset to that for the "Uniform" curves, and all data in the "Standard Deviations" area of the control panel and the worksheets are cleared. The chart curves are then blank, as they plot no data.

The distribution Option buttons select and run the distribution codes. Each control code begins by checking that the "LoadData" Check box is on, then runs all the distribution function programmes. Since each set of distribution data is located in a separate worksheet, the control code also changes the worksheet name for the chart source data.

The "Show curves" control area has 9 Check boxes to control the display of the 9 curves in the 6 charts. When a Check box is on, a curve is displayed and when it is off, the curve colour is set to that of the chart background. Each curve may thus be clearly identified.

The "Number of sets" code area enables NN to be entered either manually in the relevant cell, or rapidly changed from 100 to 1000 using a Spin button. For either case, the distribution code is run when the associated Command button is clicked. All of the 6 control codes check that the data is loaded, and the Command button checks that the appropriate distribution code is run by using logical If conditions.

Finally, the statistics in the fourth control area are derived from the separate modules. The population statistics are derived from the data module, and represent the population standard deviation (σ), and the three $\sigma/n^{0.5}$ values. The sample standard deviation statistics are computed in the distribution modules. As n or NN increase, the sample standard deviations approach the $\sigma/n^{0.5}$ values thus proving the Central Limit Theorem.

2.1.3 Theory

Example distributions

Distribution	a	b	Mean	Variance	Skewness	Kurtosis	SD
Uniform	1	6	(a+b)/2	$(b-a)^2/12$	0	1.8	1.443
Exponential	1	2	(b + a)	b^2	2	9	2
Linear	0	6	2b/3	$b^2/18$	$28b^3/135$	$60b^4/81$	1.414

Table 2.1.1: Central moments for the example distributions

Three different probability distribution functions (pdf's) are chosen to illustrate the CLT. The uniform pdf represents the equal probability of drawing from 1 to 6 digits for a sample (a six sided dice, say). The exponential pdf represents the probability of drawing more low digits (a dice weighted on the lower numbers, say). And the linear pdf represents the probability of drawing more high digits (a dice weighted on the higher numbers, say).

The pdf size and shape is usually represented by the conventional moments of the distribution, given by:

1st moment (mean)	$\mu = {}_x\int x\, f(x)\, dx$
2nd moment (variance)	$\sigma^2 = {}_x\int (x - \mu)^2 f(x)\, dx$
3rd moment (skewness)	$sk = {}_x\int (x - \mu)^3 f(x)\, dx$
4th moment (kurtosis)	$ku = {}_x\int (x - \mu)^4 f(x)\, dx$

The moment concept is adapted from Engineering Mechanics, where the same moment functions are used to analyse the strength of materials. The latter 3 moments are often referred to as central moments, since they are referenced about the first moment. (Examples of non-central moments are given in the "LMoment.xls" code description below).

Table 2.1.1 lists the 4 moments for each of the 3 example distributions. Note that the moments are, by definition, those for the entire population. Finally, the standard deviation (SD) for the population is calculated as the square root of the variance.

Example distribution functions

Distribution	pdf	cdf	idf
Uniform	1 / (b-a)	(x-a) / (b-a)	a + (b-a)F
Exponential	1/b exp[-(x-a)/b]	1 - exp[-(x-a)/b]	a - b ln(1-F)
Linear	2x/b^2	x^2 / b^2	bF^0.5

Table 2.1.2: Example Distribution functions

The pdf represents the probabability (f(x)) that a value (x) is sampled. For the uniform distribution, it is given by:

$$f(x) = 1 / (b-a) = 0.2$$

Note that the area subtended by the pdf curve evaluates to unity, since it represents the probability for all values being sampled.

The cumulative distribution function (cdf) represents the cumulative probability (F(x)) for a value not to be exceeded. For the uniform distribution:

$$F(x) = {}_a\int^x f(x)\,dx = (x-a) / (b-a)$$

The Monte Carlo method used in this code chooses n random values for F and solves for x. (This is repeated many times (NN) to find the variance of a population.) We therefore require the inverse distribution function (idf) for x. For the uniform distribution:

$$x(F) = a + (b-a)F$$

Both the cdf and idf are therefore linear curves for the uniform distribution. The other 2 example distributions are calculated similarly, and the results are given in Table 2.1.2. Using these functions, the distribution function source data was prepared using an integral range of x from 0 to 7 and calculating f(x) and F(x). This data was then entered in the "LoadData" subroutines of the "Data" module, and plotted in the "Run" charts.

Sample distributions

The sample pdf's are estimated in the "Uniform", "Exponential" and "Linear" modules of the code. The Monte Carlo method is used to generate a sample of n = 2, 4 or 9 values for F(x), and x(F) is calculated. The random samples are repeated NN times, and the frequency of the sample means (xbar) is computed. This frequency represents the non-normalised probability for a mean value, and is plotted as a pdf on the "Run" worksheet charts. (The term non-normalised is used to note the fact that the area subtended by the curve is equal to NN. It is not normalised to unity by division of NN.)

Normal distributions

Many random variables follow a Normal (or Gaussian) distribution, given by:

$$f(x) = 1/((2\pi)^{0.5} * \sigma) \exp [-(x-\mu)/2\sigma^2]$$

where μ is the mean, and σ is the standard deviation (the square root of the variance) of the population. The notation $N(\mu, \sigma^2)$ is often used to abbreviate the above expression; it was used in the summary for this code description.

The normal distributions associated with each of the f(xbar) pdf's are computed in the same modules. Firstly, the normalised $N(\mu, \sigma^2)$ is computed using the population μ and σ (Table 2.1.1), and unit intervals for the abcissa. The ordinates are then multipled by NN.

Summary of results

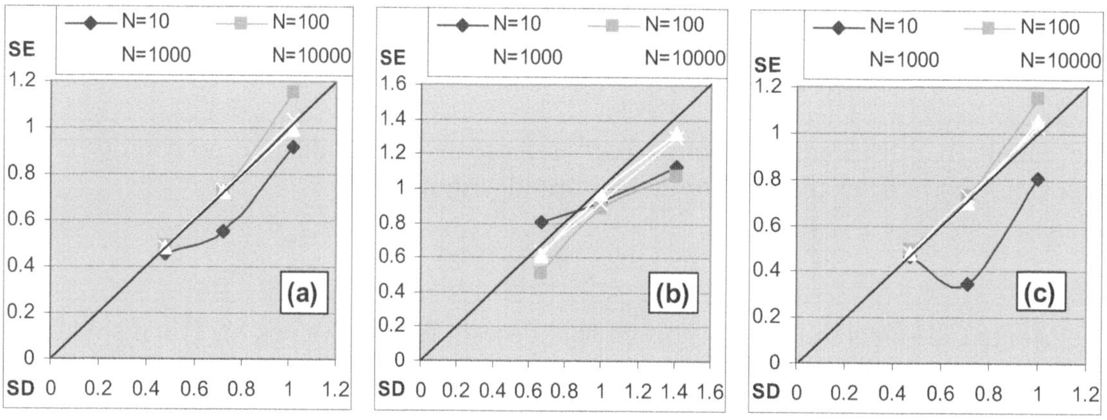

Figure 2.1.2: (a) SD of sample means versus SE of population for a uniform distribution; (b) SD of sample means versus SE of population for an exponential distribution; (c) SD of sample means versus SE of population for a linear distribution

The quotient, $\sigma/n^{0.5}$, is called the standard error of the population. Thus the CLT may be restated that as the size of the population sample (n) or the number of samples (NN) increase, the distribution of the sample means approach a normal distribution with the standard error of the population. The product of the standard error (SE) and Student's t parameter (explained in the "Regress.xls" code below) is then used as a confidence interval for the estimate of the true mean. Note that if the initial population distribution is unknown, the CLT still enables the calculation of the SE (or confidence intervals) since the SE is equal to the SD of the sample means.

An arbitrary set of results for each example distribution is shown above (Figure 2.1.2). The theoretical line, SD (sample means) = SE (population), indicates the result of the CLT. When n is increased through 2, 4 and 9 (as represented by the 3 abcissa points of each curve), most of the curves approach the theoretical line. The data is limited as a maximum of 9 points are being sampled. In practise, a sample of more than 30 points is usually necessary for true correspondence. Better correspondence is achieved as NN is increased from 10 to 10,000. In particular, the NN = 1,000 and NN = 10,000 points are close to the theoretical result.

2.1.4 Modules

There are 1 data module and 3 example distribution modules for this code. Since the distribution modules are similar, only the "Uniform" module is described herein.

Data module

The data module inputs the 3 example distributions in 3 subroutines that write data to their respective worksheets. Each subroutine is similar, and consists of 3 steps, namely:

1. A total of 13 columns are required and titled in each worksheet.

2. The distribution pdf data are written for integral x values.
3. Distribution statistics (population SD and SD/$n^{0.5}$ values) are computed and written to the "Run" worksheet.

Note that it is only these subroutines that must be edited if alternative distributions are required. If more data points are required, then the chart source data must also be edited.

Uniform module

This module consists of 2 similar subroutines for each n value, thus totalling 6 subroutines altogether. As an example, the n = 2 subroutines are "XbarUni2" and "NormUni2"; they generate the example distribution mean data and the normal distribution data respectively. The former subroutine is in 3 steps, namely:

1. Random samples are generated as described above.
2. The histogram is computed.
3. The SD of the xbar distribution is calculated and written to the "Run" worksheet.

The equivalent subroutines for higher n simply sample more values. If more than n = 9 samples are required, the sampling routine would be better coded in a loop.

The "NormUni2" subroutine calculates the normal distribution ordinates using the theoretical mean for the distribution, and the associated theoretical $\sigma/n^{0.5}$ value. The distribution is then non-normalised by multiplying the ordinates by NN. Note that these curves are better displayed non-normalised, as they illustrate the changes in NN values.

Numerical Check

The code is readily checked for accuracy by plots similar to those in the "Summary of results" given above. Note that the numerical values change each time a Command button is activated, since random data is being generated.

2.2 LINEAR AND MULTIPLE REGRESSION

2.2.1 Summary

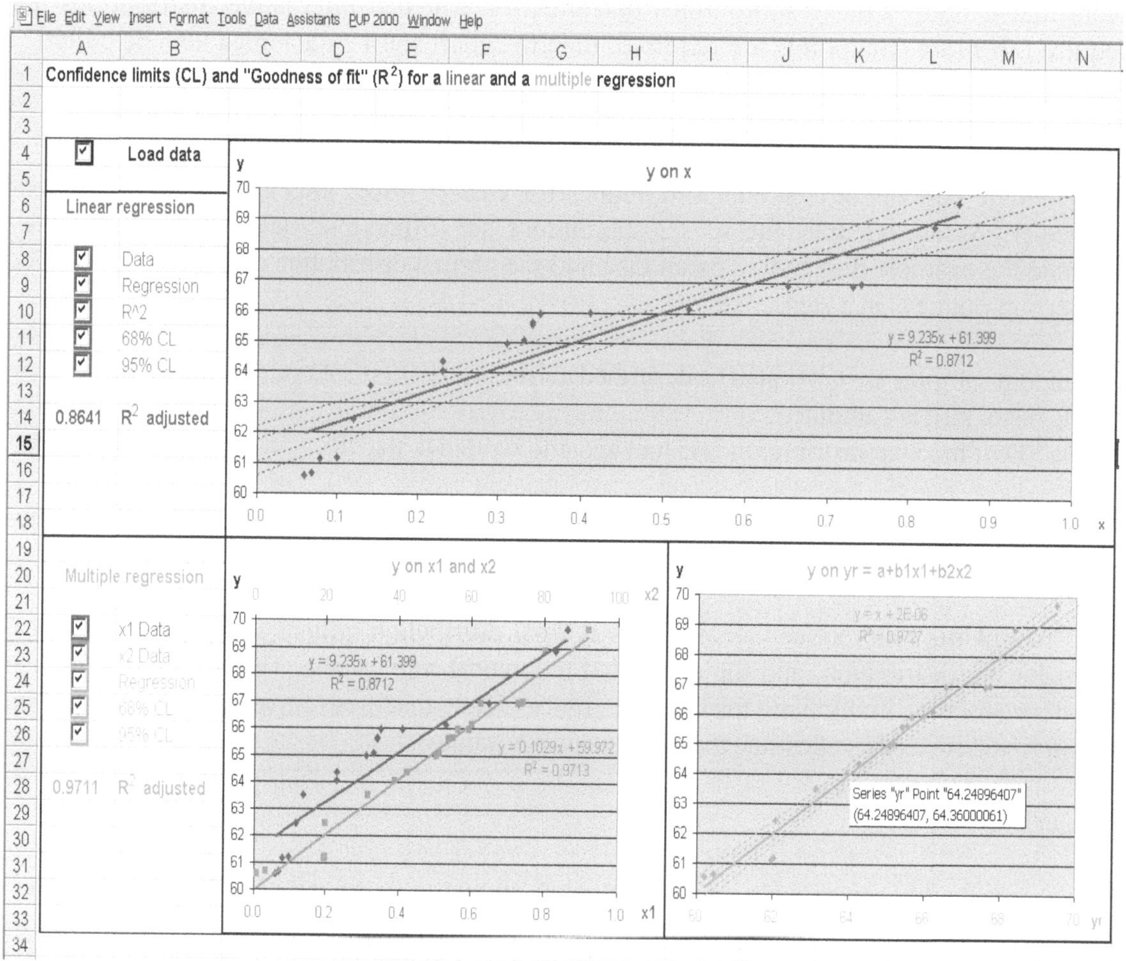

Figure 2.2.1: "Run" worksheet for the "Regress.xls" code

In hydrology, a dependent variable (y) is often expressed as a function of independent variables (x_1, x_2 ......x_n). For example, a change in reservoir storage may depend on the precipitation, evaporation, subsurface drainage and flow entering or leaving the reservoir.

In simplified examples, y may be approximately linearly dependent on a single variable ($y = a + bx$), or more variables ($y = a + b_1x_1 + b_2x_2 + \ldots$). A regression analysis is a model used to fit this linearly associated data. The x coefficients are solved to minimise the sum of the squares of the deviations in the y values, and this is called the "method of least squares". For a single x variable the method is known as linear regression, and for more than a single variable it is known as multiple linear regression (Figure 2.2.1).

2.2.2 Code operation

The code requires the operation of a single control, namely the "Load data" Check box. This control is located at the top of the control panel in the "Run" worksheet. The remaining Check boxes simply turn the chart source data on or off. The latter may be required during initial operation to identify the chart components, or to reduce the plotted information for clarity.

The "Load data" Check box is associated with the Private Sub CheckBox11_Click event in the Sheet1 (Run) object. When the Check box is clicked to On (the code is If CheckBox11.Value = True), the subroutine loads both the linear and multiple data in their respective worksheets. The two subroutines that calculate the regression coefficients and confidence limits ("LinearRegression" and "LinearCL" in the "Linear" module) are then run, and the results are written to columns in the "Linear" worksheet. The two multiple codes are also run. When the Check box is clicked to Off, all entries are cleared from the "Linear" and "Multiple" worksheets.

The Private subroutines for the remaining Check boxes (numbered 1 to 10) are also located in the Sheet1 (Run) object. Each subroutine begins with an If condition to ensure that the data has been loaded. Note that this condition is required for using alternative data from the "Load data" subroutine. A second If condition is then used to change the colour of the data plots from their True condition to the chart background colour (colour index = 15, which is 25% gray).

All the chart source data is written to the data sheets. The individual chart series may be examined by the commands:

Right click on a chart, [Source data, Series]

It is necessary to edit the data series if a regression of alternative data with more than 100 data points is required.

Note that the numerical part of this code (the "Data", "Linear" and "Multiple" modules) may be replaced by worksheet data and formulae. However, the file size is reduced with code by eliminating the storage of data. Furthermore, the use of worksheet formulae necessitates that all computed variables are stored on the worksheet. The "Multiple" worksheet then becomes large and unintelligible (since it includes large matrices).

Another reason for using a "LoadData" subroutine is that alternative data may be written to alternative "LoadData" subroutines, and saved permanently. Note that a new subroutine is best prepared by copying and editing the existing subroutine on a worksheet.

2.2.3 Theory

Linear regression

Only a brief discussion of the theory necessary for computation is given herein, and the reader is referred to Montgomery et al (2001) and Chatfield (1983) for more detail. Each dependent variable (y_i, which may be abbreviated as the "response") is related to the regressor (x_i) as follows:

$$y_i = a + bx_i + \varepsilon_i$$

where ε_i is a random error term (a residual) for a data series of say n points, and a, b are the regression coefficients. The sum of the squares of the deviations of y_i (S say) from the regression line is then:

$$S = \sum \varepsilon_i^2 = \sum (y_i - a - bx_i)^2$$

where the summation is for i = 1 to n throughout this discussion. This square error is minimised by partial differentiation of the regression coefficients, thus:

$$\partial S/\partial a = \partial S/\partial b = 0$$

The solution is then given by:

$$a = \bar{y} - b\bar{x} \qquad \text{where } \bar{y} = 1/n \sum y_i \; ; \; \bar{x} = 1/n \sum x_i$$

and

$$b = S_{xy}/S_{xx} = \sum(y_i - \bar{y})(x_i - \bar{x}) / \sum(x_i - \bar{x})^2$$

Goodness of fit

A "goodness of fit" indicator (also known as the coefficient of determination or square of the "correlation coefficient", R) is defined as:

$$R^2 = (S_{xy})^2 / [\, S_{yy} * S_{xx} \,] = [\, \sum(y_i - \bar{y})(x_i - \bar{x}) \,]^2 / [\, \sum(y_i - \bar{y})^2 * \sum(x_i - \bar{x})^2 \,]$$

Note that for a good fit, $\varepsilon_i \to 0$, $y_i \to \bar{y}$ and thus $R^2 \to 1$. For a poor fit, $\varepsilon_i \to \infty$, $(yi - \bar{y}) \to \infty$ and thus $R^2 \to 0$ since the S_{yy} term dominates the ratio. (This may be shown numerically by increasing $(y_i - \bar{y})$, and is given as an exercise for this code).

The above residual sum of squares ($S = \sum \varepsilon_i^2$) may be alternatively expressed in terms of the regressed value for y, thus:

$$S = \sum(y_i - \hat{y}_i) \qquad \text{where } \hat{y}_i = a + bx_i$$

It has been shown that, for a small n value, S has a systematic error or "bias". The unbiased estimate for S (SSE say) is then given as:

$$SSE = \sum(y_i - \hat{y}_i) / (n\text{-}p)$$

where p = 2, and is the number of regression coefficients. Similarly, an unbiased estimate for S_{yy} (SST say) is given as:

$$SST = \sum(y_i - \bar{y})^2 / (n\text{-}k)$$

where k = 1, and is the number of regressors. Using these unbiased estimates, R^2 becomes an "adjusted R^2" given by:

$$R^2_{adjusted} = 1 - SSE/SST$$

Note that $R^2_{adjusted}$ now accounts for the number of regressors used in a correlation and is therefore a better "goodness of fit" indicator.

Confidence levels

For a single random variable, it is shown in the "CLT.xls" code that the standard error (SE) for a distribution can be given as:

$$SE^2 = \sigma^2 / n$$

where σ^2 is the variance of the distribution. For a regression line, the standard error becomes a function of the abcissa value (x_0 say), thus:

$$SE^2 = SSE^2 \, [1/n + (x_0 - \bar{x})^2 / S_{xx}]$$

Note that the SE is least when $x_0 = \bar{x}$, becoming similar to that for a single variable.

The confidence interval (or limit) about a mean response ($\hat{y}_i$) is the product of the SE and a statistical value called "Student's t parameter". ("Student" was the pseudonym for the original publisher of the distribution). The t parameter represents the number of standard errors that can occur for a given percentage cumulative probability (or confidence, α). Student's t is computed in Excel using the worksheet function TINV(1-α, n-p) , thus:

$$\text{Mean response} = \hat{y}_i \pm t_{1-\alpha,n-p} * [\, SSE\, [1/n + (x_0 - \bar{x})^2 / S_{xx}]]^{0.5}$$

For a large distribution ($n \rightarrow \infty$), the t value is simply the cumulative probability of a normal distribution. Thus t = 1 for a 68% confidence interval, and t = 2 for a 95% interval. As n is decreased below about 50 however, the t value increases significantly.

Multiple regression

For a multiple linear regression, the calculation of regression coefficients and standard error reduce to the solution of a system of matrix equations. The equations are best solved by computer, since the hand solution of any matrix equation greater than 3 variables is tedious. Such computations can however be checked numerically, since the matrices may be written to a worksheet, and the results checked by substitution.

For the 2-regressor example used in this code, the response (y_i) is related to the regressors as:

$$y_i = a + bx_{i1} + b2x_{i2} + \varepsilon_i$$

This model consists of n equations that can be expressed in matrix form as:

$$\mathbf{y} = \mathbf{X\beta} + \boldsymbol{\varepsilon}$$

and can be written fully as :

$$\begin{vmatrix} y_1 \\ y_2 \\ . \\ . \\ . \\ y_n \end{vmatrix} = \begin{vmatrix} 1 & x_{11} & x_{12} \\ 1 & x_{21} & x_{22} \\ . & . & . \\ . & . & . \\ . & . & . \\ 1 & x_{n1} & x_{n2} \end{vmatrix} \begin{vmatrix} \beta_0 \\ \beta_1 \\ \beta_2 \end{vmatrix} + \begin{vmatrix} \varepsilon_1 \\ \varepsilon_2 \\ . \\ . \\ . \\ \varepsilon_n \end{vmatrix}$$

The regression coefficients are solved by minimising the sum of the squares of the residuals, and the resulting matrix solution is:

$$\hat{\boldsymbol{\beta}} = (\mathbf{X}^T\mathbf{X})^{-1}\,\mathbf{X}^T\mathbf{y}$$

where $\mathbf{X}^T$ is the transposed matrix of $\mathbf{X}$ and $(\mathbf{X}^T\mathbf{X})^{-1}$ is the inverse matrix of $(\mathbf{X}^T\mathbf{X})$. These matrix functions are readily computed in Excel using the worksheet functions [TRANSPOSE], [MMULT] and [MINVERSE]. The mean response for a particular data point is thus:

$$\hat{y}_i = \hat{\beta}_0 + \hat{\beta}_1 x_{i1} + \hat{\beta}_2 x_{i2} \qquad \text{where } \varepsilon_i = y_i - \hat{y}_i$$

The theory for R^2 follows as for the linear regression, and the standard error evaluates to:

$$SE^2 = SSE^2\,[\mathbf{x}_0^T(\mathbf{X}^T\mathbf{X})^{-1}\,\mathbf{x}_0]$$

where $\mathbf{x}_0$ represents the abcissae (x_{01} and x_{02}) for the required mean response, and is given by the matrix:

$$\mathbf{x}_0 = \begin{vmatrix} 1 \\ x_{01} \\ x_{02} \end{vmatrix}$$

The confidence interval is finally calculated as for the linear regression.

2.2.4 Modules

"Data"

The linear and multiple regression data are loaded using two subroutines that illustrate different methods of writing data to the worksheet. The "LoadDataLinear" subroutine first activates the "Linear" worksheet, titles the worksheet columns, then writes the x, y data to each cell using the VB commands:

```
Cells(2, 1).Value = 60.58: Cells(2, 2).Value = 0.06
```

The "LoadDataMultiple" subroutine declares the data as two arrays (or matrices), since these are to be used in the computations; it then writes the arrays to the "Multiple" worksheet. The arrays are, of course, the **y** and **X** matrices defined in the theory above. Note that since y and x are used as variables for the "Linear" module (see below), the arrays are given the alternative names of YY and XX; they are dimensioned in the "Declarations" section at the beginning of the module. (VB variables do not distinguish between upper and lower case, and the variable takes the case of the Declarations section.) The VB commands for the XX array is thus:

```
XX(0, 0) = 1: XX(0, 1) = 0.06: XX(0, 2) = 0.7
```

"Linear"

This module consists of 2 subroutines, "LinearRegression" for computing the regression line and "LinearCL" for the confidence limits. Although the second subroutine repeats code from the first, it is kept separate as the confidence limits are simply evaluated for 5 data points (y_i, x_i) that cover the data range.

The first subroutine is in 2 steps, namely:

1. Title and evaluate columns to compute the squared difference terms.
2. Regressed value of y (yr) to evaluate the difference sums S_{xy}, S_{xx} and S_{yy}.

The second subroutine has the following 4 steps:

1. Title and evaluate columns for the 5 selected data points.
2. Calculate the regression line slope and intercept as for the first subroutine.
3. Calculate R^2 adjusted and Student's t using the above theory for SSE and SST.
4. Apply the confidence levels using the above theory for SE.

"Multiple"

The structure of the "Multiple" module is similar to that of the "Linear" module. The steps for the "MultipleRegression" subroutine are:

1. Evaluate matrices by transposition and multiplication to give BETA (**β**).

2. Calculate the regressed value of y (yr) using the BETA values.

The steps for the "MultipleCL" subroutine, which again uses only 5 representative data points, are:

1. Title the columns similar to the "Linear" worksheet columns.
2. Choose arbitrary data covering the variable ranges (new arrays, X1(0,2) to X5(0.2)).
3. Calculate the regressed value of y (yr) for each arbitrary data point.
4. Calculate the variance of yr using the above matrix theory.
5. Calculate the confidence limits as for the linear regression.

Numerical Check

A general check can be made by ensuring the data appears randomly scattered about the regression line. The confidence limits should increase outside the data range, and $R^2_{adjusted}$ should be less than the regression R^2. For the multiple regression line, R^2 for the combined data should be higher than for the individual data and the slope of "y on yr" should be unity. A numerical check is however still necessary. The first data line of the "Linear" worksheet may be checked as follows:

Column A: 60.58
Column B: 0.06
Column C: Evaluate xm = [WorksheetFunction.Average(B:B)] = 0.37; x-xm = -0.31
Column D: $(x\text{-}xm)^2 = 0.096$
Column E: Evaluate ym = [WorksheetFunction.Average(A:A)] = 64.84; y-ym = -4.26
Column F: $(y\text{-}ym)^2 = 18.15$
Column G: (x-xm)(y-ym) = -0.31*-4.26 = 1.32
Column H: Sxx and Sxy can be evaluated using [WorksheetFunction.Sum]; yr = 61.95
Column I: (y-yr)^2 = -1.37*-1.37 = 1.88

The result for Column I is similar to that computed, with a small error due to rounding of the data. The confidence limits may be finally approximated to ± SE and ± 2*SE where:

SE = (SSE * (1 / n + (Cells(i, 11).Value - xm) ^ 2 / Sxx)) ^ 0.5 = 0.38 for i = 2

The multiple regression may be checked in a similar manner, by writing the matrices to the worksheet.

2.3 EXTREME VALUE ANALYSIS

2.3.1 Summary

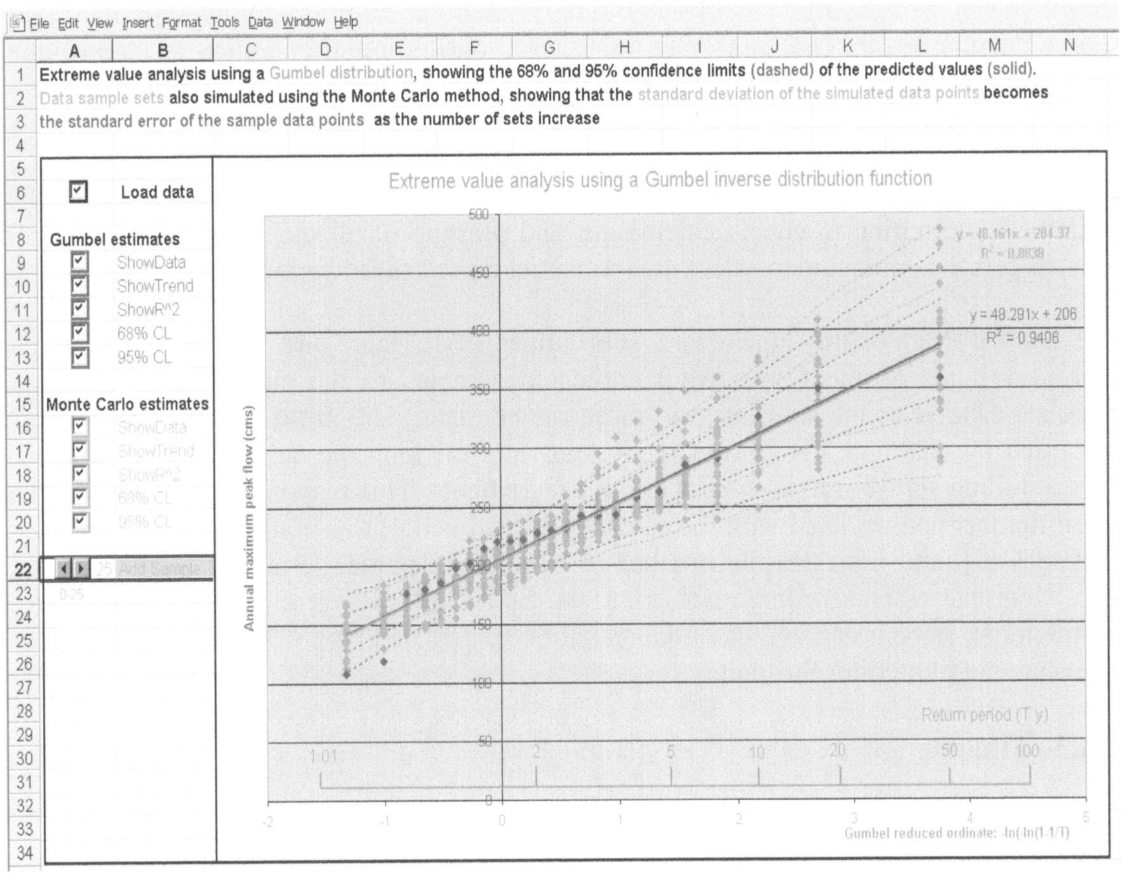

Figure 2.3.1: "Run" worksheet for the "EVA.xls" code

As an example for extreme value analysis (EVA), we consider the forecasting of flood events. Statistical methods are used to assign a probability to a particular flood magnitude. A hydraulic structure may thus be designed, with known risk, to withstand such a flood event. In particular, we analyse the annual maximum flows (in m^3/s) for 24 years of data from a continuously gauged river record using the Gumbel frequency distribution (Figure 2.3.1). The resulting inverse distribution function enables the estimate for a flood magnitude (Q m^3/s) that may be equalled or exceeded once in every T years, where T is called the return period.

The standard error (SE) for this estimate is also calculated from the theory; it is plotted as a 68% confidence limit (approximately Q ± SE) and a 95% confidence limit (approximately Q ± 2SE). The Monte Carlo method is then used to simulate an additional number of data sets (j). The standard deviation for the simulated data is shown to be equal to the theoretical standard error for the sample points as j is increased. The simulation therefore offers a non-theoretical method for computing a confidence limit.

2.3.2 Code operation

This code requires the operation of 2 controls only, namely the "Load Data" Check box and the "Add Sample" Spin Button. The latter are located at the top and bottom of the control panel in the "Run" worksheet. The remaining controls simply turn the chart source data on or off. The latter may be required during initial operation to identify the chart components, or to reduce the plotted information for clarity.

In accord with the operation of all the statistics programmes, the "Load Data" Check box must be operated first. This writes the data to the "Gumbel" worksheet and runs the "Gumbel" subroutine to enable calculation and plotting of all the associated curves. It also triggers the command to select the sequential Spin button control.

The Spin button control operates 2 subroutines, "MCData" and "MCPlot". The first subroutine adds Monte Carlo (MC) simulated samples to the chart, and the second calculates the relevant statistics for these added data. The Spin button [Properties] (accessed by enabling VB Design mode and right clicking the activated Spin button) have scrolling set from zero to 26 with unit increments. This may be altered at will. The scroll number is associated with the same worksheet cell as the Spin Button [Properties, Linked Cell], and indicates the number of simulated samples (j) added to the original data. Note that when scrolling reaches 26, the Spin button control code resets j to zero. Furthermore, since a standard deviation cannot be calculated for $j < 3$, thc MC confidence limits are not plotted for this range.

2.3.3 Theory

Introduction

At least 3 assumptions are implied in the data used for the EVA of flood design, namely:

1. The data is random and thus all events are independent. For an annual series, it may happen that a peak flow occurs in the same winter for consecutive years and are dependent (i.e. they occur for the same storm). It is therefore better to define an annual series in terms of Water years instead of Calendar years.
2. The natural processes are stationary with time (i.e. the series moving average is constant). Examples of non-stationary processes are Global warming (long term) and structural changes such as dams (short term).
3. The population parameters may be estimated from the sample. This implies that the data record must be acquired over sufficient time as to estimate the risk associated with the same time. Any extrapolation is subjective.

The other main assumption used in EVA is the type of frequency distribution used. In the past, most types were selected so as to enable a straight line to be fitted through the data. With the advent of modern computers however, this is no longer necessary and a linear curve with minimal data scatter is desired. There remain many choices for a probability distribution function. Past examples have been given by Kite (1977). He details the theory for at least 10 types that use conventional moment analysis. The associated computer codes for these types, written in Fortran IV, are also given. More recent examples have been given by Hosking and Wallis (1997); they detail 11 types or groups with up to 5 variables, and which use linear moment analysis. Codes for their enumeration are given in Fortran 90 (http://lib.stat.cmu.edu/general/lmoments).

A commonly used frequency distribution is that proposed by Gumbel (1958). It is used because the abcissa is plotted as a double logarithm of the probability function. The double logarithm tends to spread apart the plot of high extreme values, and thus enables the assumption of a straight-line prediction. The distribution is used herein as a single example for EVA. The discussion follows that given by Shaw (1994), and uses similar nomenclature for ease of reference.

Probability density function

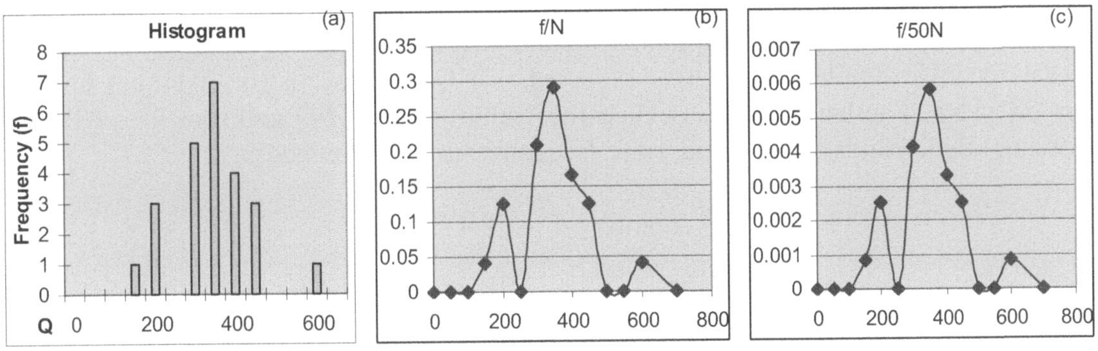

Figure 2.3.2: (a) Histogram of data (b) Frequency curve with non-dimensionalised ordinate (c) Normalised frequency curve

A histogram of the example peak flow data used in this code may be plotted using the [Column] chart. Figure 2.3.2(a) shows the frequency or number of occurrences (f) of annual peak flows (Q) within a class interval (ΔQ) of say 50 m^3/s. The frequency ordinate for say n classes may be non-dimensionalised by dividing by the number of data (N = 24) so that ${}_0\sum^n f/N = 1$ (Figure 2.3.2(b)). Furthermore, the curve may be normalised (i.e. its subtended area is unity) by further dividing f/N by ΔQ (Figure 2.3.2(c)), thus:

$$\text{Subtended area} = {}_0\int^{\infty} f/(N\,\Delta Q)\,dQ = {}_0\sum^n f/N = 1$$

As ΔQ→0 and N →∞, the curve will become smooth. The normalised ordinate then represents the probability (p) of observing a Q value, and the curve is called the "probability distribution function" (pdf). Three examples of discrete pdf's are given in the "CLT.xls" code. The Gumbel pdf is however more complex and is given by Kite (1977) as:

pdf: $$p(x) = b \exp[-b(x-a)-\exp(-b(x-a))]$$

The 2 parameters a and b are solved from the first and second moments of this pdf, thus:

$$a = \overline{x} - 0.5772/b \qquad \text{and} \qquad b = \pi / s\sqrt{6}$$

where $\overline{x}$ (the mean value) and s^2 (the variance) are the first and second moments. For the sample used herein, they are estimated as:

$$\overline{x} = \overline{Q} = 1/N * {}_i\sum Q_i \qquad \text{and} \qquad s^2 = 1/(N-1) * {}_i\sum (Q_i - \overline{Q})^2$$

The Gumbel pdf can thus be solved analytically. It may, for example, be integrated to find the probability (P) that an annual maximum flood (say Q_p) is equalled or exceeded, thus:

$$P = {}_{Qp}\int^{\infty} p(Q)\, dQ$$

It follows that the probability of non-exceedence, F = 1-P, is given by the "cumulative distribution function" (cdf) as explained for the "CLT.xls" code. If say P = 1/24 for the 24 year record example herein, it is expected that Q_p will occur (on average) once in every 24 years. The latter value is called the "return period" (T) and thus T = 1/P. The following relationships for P, F and T are frequently used in the theory:

$$T = 1/(1-F) \qquad \text{and} \qquad F = (T-1)/T$$

Inverse distribution function (idf)

The ultimate goal for EVA is to determine a design flood for a given return period. This requires the enumeration of the idf, which is given for this example as Q = f (T). The idf is derived by inverting the cdf, and the latter is derived by integrating the pdf, thus:

cdf: $F(x) = \exp[-\exp(-b(x-a))]$

idf: $x = a - 1/b\ \ln(-\ln F) = a - 1/b\ \ln(-\ln(1-1/T))$

The function $-\ln(-\ln F)$ is called the "Gumbel reduced variate", and the idf should plot as a straight line if the distribution is true. Predicted values for x ($\hat{x}$) will fall on this line. The present example for 24 data values (worksheet "Run") give the idf as:

idf: $\hat{x} = 206 + 48.291\ (-\ln(-\ln(1-1/T))$

The idf be may be alternatively expressed in terms of the mean and variance using the above relations from the pdf moments, thus:

idf: $\hat{x} = \overline{x} - \sqrt{6}s/\pi\ (0.5772 + \ln(-\ln(1-1/T)) = \overline{x} + K(T)s$

The function K(T) is called the "Frequency factor", and simply depends on T (and thus F or P).

The standard error (SE) for a distribution is defined as:

$$SE^2 = 1/N \, {}_i\Sigma \, (x_i - \hat{x})^2$$

It has been shown theoretically by Kite (1977) that for the Gumbel, 2-parameter distribution, the SE becomes:

$$SE^2 = s^2/N \, [1 + 1.14K(T) + 1.10(K(T))^2]$$

where s is the standard deviation of the sample (i.e. the square root of the variance, s^2).

Finally, two confidence limits for the distribution, 68% and 95%, are estimated using Student's t distribution (Montgomery et al, 2001). The parameter $t_{\alpha,v}$ is computed using the Excel "TINV" worksheet function, where (1-α) is the confidence limit and ν is the degree of freedom (N-1)). Thus an estimate for a T year design flood is given as $\hat{Q} \pm t_{\alpha,v} * SE$.

Monte Carlo simulation

The Monte Carlo method simulates the entire population from which the sample of 24 data points is derived. As the number of simulated samples proceed to infinity, the standard deviation of the mean value for each data point ($\hat{x}$) approaches the standard error of the original sample (by reason of the SE definition above). The code demonstrates this for up to 25 simulated samples, and may be extended for more samples.

The idf equation from worksheet "Run", given above, is used to simulate a sample:

idf: $\hat{Q} = 206 + 48.291 \, (-\ln(-\ln F))$

Random samples of F (where $0 < F < 1$ and each sample is 24 points) are generated to give simulated sets of 24 values for Q. Each set is sorted and ranked as for the original data, and therefore assigned the same Gumbel reduced ordinate. The code is then used to calculate the mean and standard deviation for Q at each Gumbel ordinate. (Note that the simulation is effectively applying the Central Limit Theorem to each Gumbel ordinate. The "CLT.xls" code was used to show that the mean of a number of samples approached a normal distribution, whereas it is now extended to any ordered statistic of a number of samples.)

2.3.4 Modules

Only one "Data" module is used for this code, and it includes 4 subroutines. The "LoadData" subroutine is common to other codes, and is best edited by copying to a worksheet. The remaining subroutines calculate the Gumbel, Monte Carlo data and Monte Carlo statistics respectively.

The "Gumbel" and "MCPlot" subroutines reduce to fairly simple operations on each column of data. They may therefore be alternatively programmed using worksheet formulae. They are however programmed in VB herein to reduce the size of the "EVA.xls" file. (Note that VBA can use 2 types of mathematical functions, namely intrinsic functions such as "ABS" or "RND" and worksheet functions such as "AVERAGE" and "STDEV". The list of available functions are obtained from the VB "Help menu" by entering the Index keywords "Math function" or "Worksheet function".)

"Gumbel" subroutine

This subroutine writes data to the columns of the "Gumbel" worksheet, using a single loop. The data statistics of average (Qbar), standard deviation (s) and number (n), and the Students's t values (ST68, ST95) are first computed. The columns follow:

Column A: original sorted data of n records written from the "LoadData" subroutine
Column B: rank of the data (r), given by the loop variable (i)
Column C: Gringorten probability of exceedence, P = (r-0.44)/(n+0.12), which becomes equal to r/n for large r and n (Shaw, 1994)
Column D: Return period, T = 1/P
Column E: Gumbel reduced variate = -ln(-ln(1-1/T))
Column F: frequency factor, K(T) given above
Column G: predicted data, $\hat{Q} = \bar{Q} + K(T)*s$
Column H: standard error, $SE = s/\sqrt{n}\,[1 + 1.14K(T) + 1.10(K(T))^2]^{0.5}$
Column I: upper 68% confidence limit (CL) = $\hat{Q}$ + ST68*SE
Column J: lower 68% confidence limit (CL) = $\hat{Q}$ - ST68*SE
Column K: upper 95% confidence limit (CL) = $\hat{Q}$ + ST95*SE
Column L: lower 95% confidence limit (CL) = $\hat{Q}$ - ST95*SE

"MCData" subroutine

This subroutine consists of 2 steps, namely:

1. The random series is generated using the rnd function. Note that the rnd function is multiplied by 1000, to obtain 24 random integers between 0 and 1000, then divided by 1000 to obtain 3 decimal places for F. Furthermore, F cannot equal zero or unity on a logarithmic scale, so the upper and lower bounds for rnd are restricted to 999 and 1. The generated series are written in sequential columns to the "MCData" worksheet.

2. The MCData are sorted in descending order and written in sequential rows to Column A of the "MCPlot" worksheet. The corresponding Gumbel reduced variates are written to column B. The total data is then available for plotting on the worksheet "Run" chart.

"MCPlot" subroutine

This subroutine begins with the condition to end the code if the number of generated data sets < 3 (since the standard deviation cannot be computed). The code is then similar to the "Gumbel" subroutine, as a single loop is used to generate the columns for the "MCPlot" worksheet:

Column C: Gumbel reduced variate = -ln (-ln (1-1/T)) for a single series
Column D: average of generated data for each reduced variate
Column E: standard deviation (SD) of generated data for each reduced variate
Column F: upper 68% confidence limit (CL) = $\bar{Q}$ + 1*SD
Column G: lower 68% confidence limit (CL) = $\bar{Q}$ - 1*SD
Column H: upper 95% confidence limit (CL) = $\bar{Q}$ + 2*SD
Column I : lower 95% confidence limit (CL) = $\bar{Q}$ - 2*SD

Numerical Check

The convergence of the sample and Monte Carlo confidence limits provide a numerical check for the computation. However, a numerical check on say, the first lines of the "Gumbel" and "MCPlot" worksheets ensures familiarity with the method. For example, the first data row of the "Gumbel" worksheet is checked as follows:

Column B: 1
Column C: P = (r-0.44)/(n+0.12) = 0.0232
Column D: Return period, T = 1/P = 43.07
Column E: Gumbel reduced variate = -ln(-ln(1-1/T)) = -ln(-ln0.0976) = 3.75
Column F: K(T) = $\sqrt{6}$/3.142(-0.5772 - ln(-ln(1-1/T) = 0.7795(-0.5772 + 3.75) = 2.475
Column G: $\hat{Q} = \bar{Q} + K(T)*s$ = 233.04 + 2.475*61.11 = 384.28 cms
Column H: SE = $s/\sqrt{n}\,[1 + 1.14K(T) + 1.10(K(T))^2]^{0.5}$ = $61.11/4.90(1+2.82+6.74)^{0.5}$
= 40.52 cms

The confidence limits may be finally approximated to ± SE and ± 2*SE.

Alternative data

The inclusion of alternative data in the "LoadData" subroutine has been discussed. Note that an "Excel 2000" worksheet can hold up to 65,536 rows of data. A data limit is thus set by the "MCPlot" worksheet since it records all the generated data. Another change required for alternative data is that of inserting a new trendline equation in the "MCData" subroutine (constants m and c). Futhermore, the source data for the chart in worksheet "Run" has been set to plot n = 100, and must be edited for n > 100.

2.4 LINEAR MOMENT ANALYSIS

2.4.1 Summary

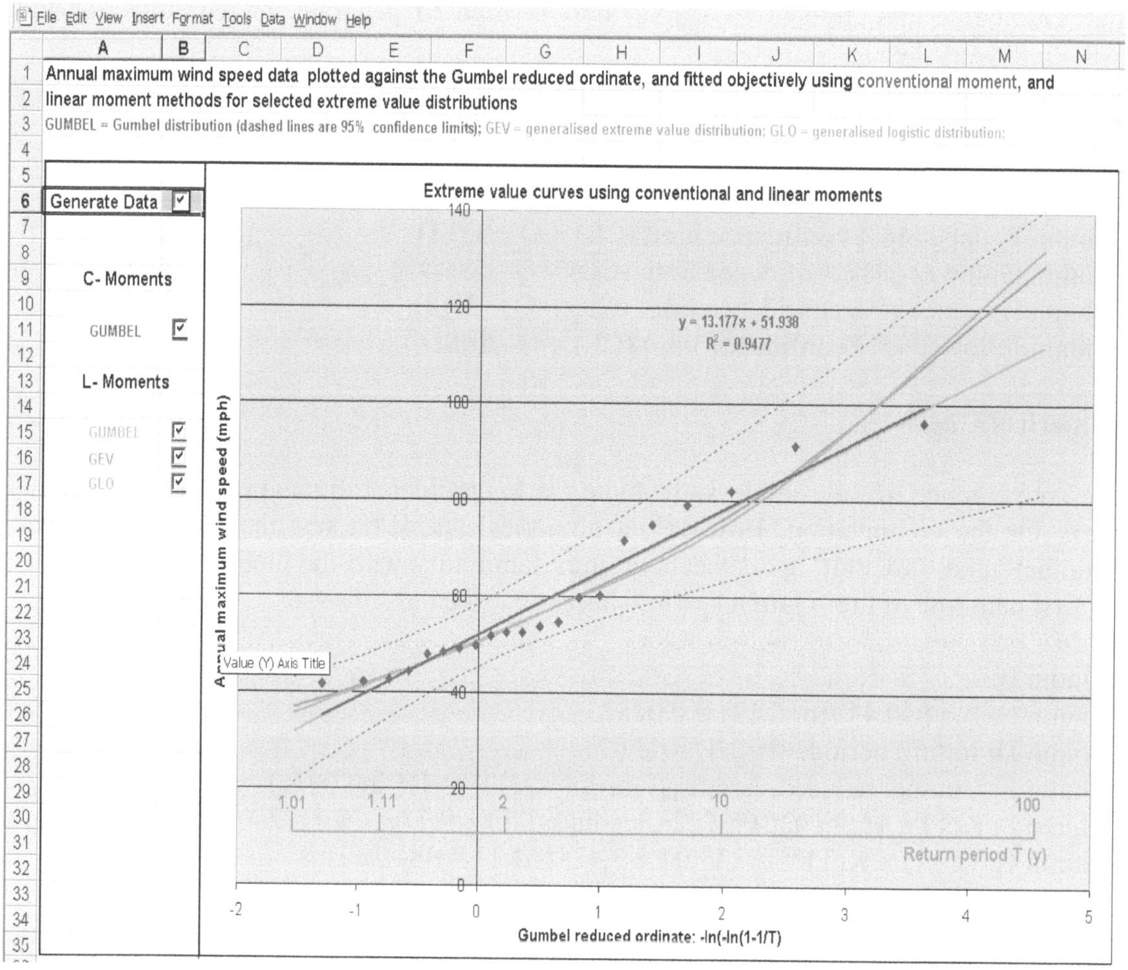

Figure 2.4.1: "Run" worksheet for the "LMoment.xls" code

The "EVA.xls" project examined extreme values of river flood discharges for a single type of frequency distribution. This present code examines extreme wind speeds for a similar distribution using both convential moment analysis (C-Moments) and a more recent, linear moment analysis (L-Moments, Hosking and Wallis, 1997). Linear statistics have the advantage of being able to simulate a wider variety of frequency distributions, and are better able to follow objectively the presence of outliers in the data. The latter is useful for estimating the return periods (T) for both extreme high and extreme low events.

The random examples contrast extreme value curves for a Gumbel distribution using C-Moments. The Gumbel, Generalised Extreme Value (GEV) and Generalised Logistic distribution are also computed using L-Moments, to show a better representation of data outliers (Figure 2.4.1).

2.4.2 Code operation

There are 3 sets of controls in the control panel, namely the "Generate data" Check box, the "C-Moments" Check box, and the 3 "L-Moments" Check boxes. The "Generate data" control must be true (ticked) for any of the remaining controls to operate. It operates the entire project, by loading a random data set and calculating all the statistics. The "C-Moments" and "L-Moments" Check boxes are present to turn the associated curves on of off, thus enabling identification of the curves.

The plotted C-moment data on the "Run" worksheet consists of the data points, a linear trend analysis through the data using the Gumbel distribution, and the 95% confidence limits (dashed lines). The L-moment data for the 3 distributions are shown in different colours, and the improved fitting of the GEV and GLO distributions are apparent. Note that confidence limits can be included for these L-Moment distributions by trending the curves with a polynomial fit, and generating Monte Carlo samples as for the "EVA.xls" code.

2.4.3 Theory

Discrete C-Moments for a sample

The conventional mean for a data sample $x_1, x_2 \ldots\ldots x_n$ is given by its first moment:

$$\bar{x} = 1/n \; {}_{i=1}\sum^{n} x_i$$

where $\bar{x}$ is an estimate for μ, the population mean or "location" of the data (Figure 2.4.2(a)). The variance is the second moment of the sample, thus:

$$s^2 = 1/n \; {}_{i=1}\sum^{n} (x_i - \bar{x})^2$$

where s^2 is an estimate for σ^2, the population variance or the square of the standard deviation (alternatively known as the "scale" or spread of the data, Figure 2.4.2(b)).

In general, C-Moments have higher sample moments given by:

$$m_r = 1/n \; {}_{i=1}\sum^{n} (x_i - \bar{x})^r$$

The 3rd and 4th moments may be used to estimate the "skewness" (symmetry, Figure 2.4.2(c)) and "kurtosis" (peakedness, Figure 2.4.2(d)) of the population distribution.

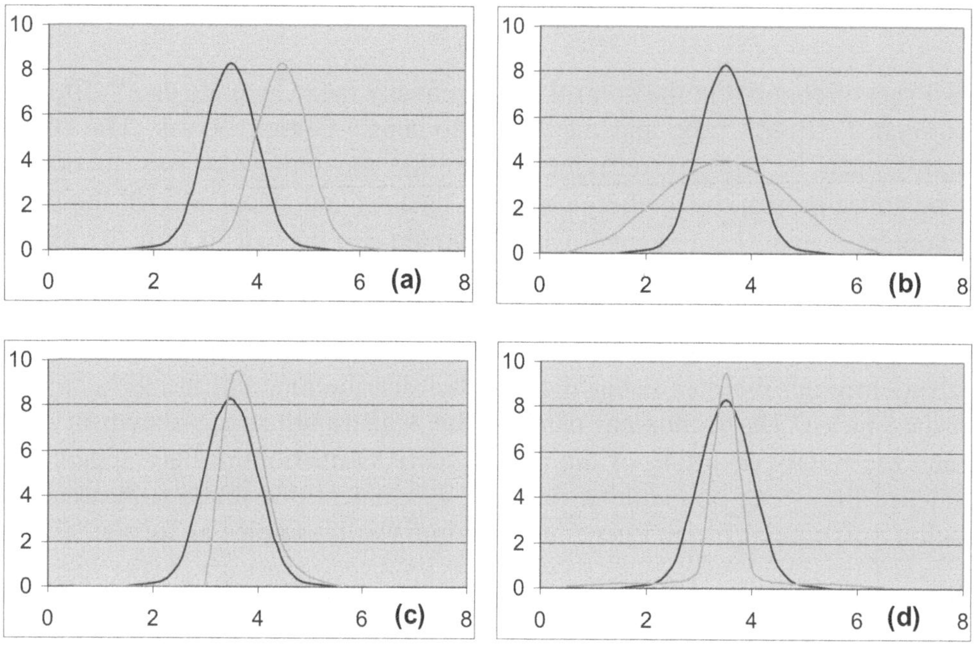

Figure 2.4.2: (a) 2 positions for the location (mean); (b) 2 values for the scale (standard deviation); (c) Skewness of a distribution (symmetry); (d) Kurtosis of a distribution (peakedness)

It has been frequently shown that C-Moment estimators have associated systematic errors or "bias" (Hosking and Wallis, 1997). For example, the unbiased estimate of s^2 is calculated by replacing n with (n-1), thus:

$$s^2 = 1/(n-1)\ {}_{i=1}\sum{}^{n}\ (\overline{x}_i - x)^2$$

Estimates based on higher moments are more unreliable. For this reason, the L-Moment method has been researched, and a better set of unbiased results established.

Continuous C-Moments and L-Moments for a population

The cumulative distribution function (cdf), F(x), becomes a continuous curve for large n. It defines the probability (P) that any value X is not exceeded, thus:

$$P(x \leq X) = {}_{-\infty}\int^{X} F(x)\ dx$$

The population mean or first moment of the distribution is given by:

$$\mu = E(X) = {}_{-\infty}\int^{+\infty} x\ dF(x)$$

where E(X) is called the "expected value" of the random variable X (Stuart and Ord, 1987), and is an abbreviation for the integration moment. Thus E(X) is the area subtended by the inverse distribution function , x(F). As an example, we can take the area subtended by x(F) for the uniform distribution given in the "CLT.xls" code:

$$\text{Subtended area} = \mu = 0.5\ (5 * 1) + (1 * 1) = 3.5$$

If we arbitrarily substitute the transformation u = F(x), then E(x) is more simply written :

$$\mu = E(X) = {}_0\int^1 x(u)\ du$$

since $0 < F(x) < 1$. The higher moments may then be written as:

$$E(X^r) = {}_0\int^1 [x(u)]^r\ du$$

L-Moments were derived from "probability weighted moments" of a random variable, X. For an idf of x(u), the principal L-Moments are given by:

$$E(X^r) = {}_0\int^1 x(u) * (u)^r\ du$$

L-Moments are thus integrals of the first order in x(u), and weighted by the polynomials u^r . They are therefore simpler to compute than C-Moments.

Discrete L-Moments for a sample

To estimate L-Moments, the sample data is first sorted in ascending order. For a data set of n values, the ordered series is thus given by $X_{1:n} \leq X_{2:n} \leq \dots\dots\dots\dots \leq X_{n:n}$. The computations begin with the estimate of the "probability weighted moments", b_r (B-Moments). The first 4 B-Moments are given by:

$$b_0 = 1/n \ {}_{j=1}\sum^n X_{j:n}$$

$$b_1 = 1/n \ {}_{j=2}\sum^n (j-1)/(n-1) * X_{j:n}$$

$$b_2 = 1/n \ {}_{j=3}\sum^n [(j-1)(j-2)] / [(n-1)(n-2)] * X_{j:n}$$

$$b_3 = 1/n \ {}_{j=4}\sum^n [(j-1)(j-2)(j-3)] / [(n-1)(n-2)(n-3)] * X_{j:n}$$

The L in L-Moments emphasises their derivation from linear combinations of order statistics. Sample L-Moments, l_r, are thus derived linearly from the B-Moments as:

$$l_1 = b_0 \ ; \ l_2 = 2b_1 - b_0$$

$$l_3 = 6b_2 - 6b_1 + b_0 \ ; \ l_4 = 20b_3 - 30b_2 + 12b_1 - b_0$$

Since L-Moment ratios (t_r) frequently occur in frequency distribution analysis, they are defined as follows:

$$t_3 = l_3/l_2 \; ; \; t_4 = l_4/l_2$$

Hosking and Wallis (1997) detail 11 types or groups of distribution functions, and solve their pdf, cdf and idf in terms of l_1, l_2, t_3 and t_4 only. Three common types are illustrated in this present code.

2.4.4 Modules

Two modules are used for this code, namely the "Data" module and the "Quantiles" module. The term "Quantile" is used as an alternative name for the idf ordinate, x(F), and denotes selected values for x(F). The idf may thus be alternatively called the "quantile function".

"Data"

There are 2 alternative subroutines to load data. The first subroutine, "GenerateData", writes a random series of trended data to Column A of the "Design" worksheet. This data changes whenever the "Generate Data" Check box is activated. An alternative and constant data set may also be loaded using the "LoadAlternativeData" subroutine. This data set is loaded by changing the macro commands in the "Generate Data" Check box control code.

The load data subroutines are followed by a "Plotdata" subroutine with 2 steps. The first is to re-sort the ascending ordered data in the "Design" worksheet (Column A) to descending order (Column F). The second step is to compute the Gumbel statistics (Columns B to E) for plotting in the "Run" worksheet.

"Quantiles"

This module consists principally of 4 subroutines that calculate the extreme value data (x(F)) for 5 quantiles which represent return periods of 1.01, 1.11, 2, 10 and 100 years. The 4 subroutines are "Gumbel" for C-Moments, and "GUMQuantiles", "GEVQuantiles" and "GLOquantiles" for the Gumbel, Generalised extreme value and Generalised logistic distributions using C-Moments. A 5th subroutine, "LMoments", computes the L-Moments for the data of each station.

The "Gumbel" subroutine differs slightly from that in the previous "EVA.xls" code, since the curve is computed for only 5 quantiles. The extreme values are simply calculated in sequence, rather than the looping used in the previous code. Nevertheless, the mathematics is the same and can be readily inferred from the previous code.

The "LMoments" subroutine computes l_1, l_2, t_3 and t_4 in 4 steps, after counting the data and naming the worksheet columns in "Design". The calculations follow the B-Moment and L-Moment formulae given above, and the L-Moments and L-Ratios are written to the "Design" worksheet for access by the remaining subroutines. Note that l_1 is simply the average of the data as used in C-Moments. The remaining B-Moment summations are computed by calculating the first value (a) and looping through the variable "Sum = Sum + a". L-Moment linear additions are then applied.

The L-Moment quantile subroutines compute x(F) sequentially, and use the following distribution and parameter values from Hosking and Wallis (1997).

Gumbel 2 parameter : $x(F) = \xi - \alpha \ln(-\ln F)$

ξ (the location) $= l_1 - 0.5772\alpha$; α (the scale) $= l_2 / \ln 2$

GEV 3 parameter: $x(F) = \xi + \alpha [1 - (-\ln F)^k] / k$

$\xi = l_1 - \alpha [1 - \Gamma(1+k)] / k$; $\alpha = l_2 k / [(1-2^{-k}) \Gamma(1+k)]$;

k (the shape) $\approx 7.8590c + 2.9554c^2$ where $c = 2/(3+t_3) - \ln 2/\ln 3$

GLO 3 parameter: $x(F) = \xi + \alpha [1 - ((1- F) / F)^k] / k$

$\xi = l_1 - \alpha [1/k - \pi / \sin k\pi)]$; $\alpha = l_2 \sin k\pi / k\pi$; $k = t_3$

Numerical Check

A general check can be made on the L-Moment predictions, since the curves should give greater weight to the extreme tails of the distribution. Particular checks can however be made for the quantiles of each distribution. The example below gives a check for the first quantile (T = 1.01) of the data from the "LoadAlternativeData" subroutine, using the Gumbel distribution with L-moments.

1. n = 28; T = 1.0101; F = 1 –1/1.0101 = 0.01; -ln(-ln0.01) = -1.53
2. $b_0 = 1/28 \; {}_{j=1}\sum^{28} X_{j:n} = 1261/28 = 45.036$
3. $b1 = 1/28 \; {}_{j=2}\sum^{28} (j-1)/(n-1)X_{j:n} = 1/28 \; (1/27*X_2 + 2/27*X_3 + \ldots 27/27*X_{28}) = 373.190/28 = 24.746$
4. $l_1 = 45.036$; $l_2 = 2b_1 - b_0 = 2*24.746 - 45.036 = 4.456$
5. $\alpha = l_2/\ln 2 = 4.456/0.693 = 6.429$
6. $\xi = l_1 - 0.5772\alpha = 45.036 - 0.5772*6.429 = 41.325$
7. $x(0.01) = \xi - \alpha \ln(-\ln F) = 41.325 + 6.429*(-1.53) = 31.5$

This quantile agrees with the computed data in the "Design" worksheet. Note that the separate moments may be checked by debugging the code (using the VB Debug and Run to Cursor commands), and placing the cursor on a required variable.

2.5 ORDINARY AND UNIVERSAL KRIGING

2.5.1 Summary

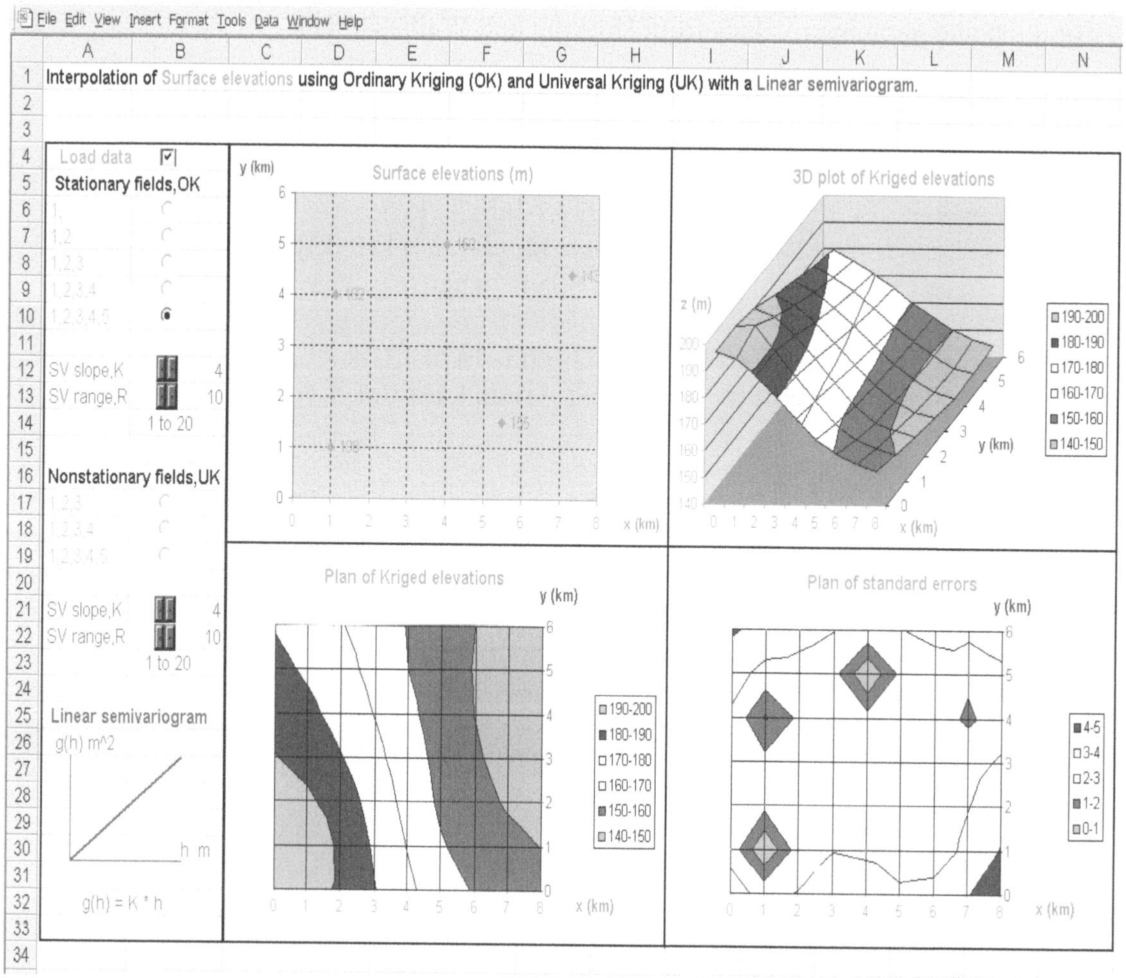

Figure 2.5.1: "Run" worksheet for the "OKrige.xls" code

Kriging is named after a South African mining engineer (Krige, 1966) and was also developed at the Centre de Morphologie Mathematique in France in the 1960's (Matheron, 1963). The historical development of the method is summarised, for example, in Webster and Oliver (2001). It is a statistical method for interpolating between known data points, and is particularly useful in contouring surfaces. The example project herein is for the contouring of ground water elevations (z) measured at several points (x, y). From 1 to 5 data points may be selected, and Kriging estimates the contour elevations with the associated standard errors (Figure 2.5.1). Although other methods exist to contour point elevations (such as inverse distance weighting), the distinguishing feature of Kriging is that it minimises the estimated standard errors (Isaaks and Srivastava, 1989).

2.5.2 Code operation

The data points are selected by means of the Option buttons presented. Since the ground water surface is considered continuous between each data point, it must be spatially correlated over a short distance known as the range (R). This correlation is measured by the semivariogram (SV) of a regionalised variable (Davis, 1986, and explained later in the "GWlevels.xls" code). A linear SV is assumed for this example as shown in the control panel of the "Run" worksheet. The SV slope (K) may be varied from 1 to 20 m^2/km and the range (R) from 1 to 20 km by means of the 2 Spin buttons.

Ordinary Kriging, as illustrated above, is no longer relevant if there is a slow change in the average value of the regionalised variable (i.e. it is no longer stationary and $\partial z/\partial x \neq 0$, $\partial z/\partial y \neq 0$ for ∂x and $\partial y > R$). A slow change may consist of a non-stationary trend and possibly stationary residuals; its solution is called Universal Kriging (Bras et al, 1993). The simplest linear trend surface is described by 3 coefficients ($z = b_0 + b_1x + b_2y$). Universal Kriging for 3 data points therefore gives this trend surface. As the data points are increased to 5, Ordinary Kriging is solved for the stationary residuals. The semivariogram slope and range may be varied as for Ordinary Kriging by means of the 2 Spin buttons.

For both Ordinary and Universal Kriging, the physical effect of increasing the SV slope means there are larger differences in the regionalised variable over the same distance between points, thus the standard error of the estimated contour will increase. Correspondingly, the physical effect of increasing the range means that more contour points influence each other in the solution. For example, the Universal Kriging for 5 stations at a 1 km range is equivalent to a linear trend, since all points are more than this distance apart.

2.5.3 Ordinary Kriging for 2 stations

Theory

The ultimate object is to determine a surface from known points (z_i) with minimum standard error (SE). This surface can be mapped for a grid of points (z_p) using the surface-plotting feature of Excel charts. A grid point can be estimated as a weighted average of the known points, thus:

$$\hat{z}_p = \sum W_i z_i$$

And the variance, $(SE)^2$ follows as:

$$(SE)^2 = 1/n \sum (\hat{z}_p - z_p)^2$$

The Kriging equations (Olea, 1975) were derived to solve W_i for a minimum SE by differentiation. For only 2 stations, the weights can be reduced to 2 simultaneous equations, as follows:

$$W_1 g(h_{11}) + W_2 g(h_{12}) = g(h_{1p})$$
$$W_1 g(h_{12}) + W_2 g(h_{22}) = g(h_{2p})$$

where $g(h_{ij})$ is the semivariance for a distance h between points z_i and z_j, and must be calculated before Kriging. A third equation constrains the weights to sum to unity, thus

$$W_1 + W_2 = 1$$

Since there are 3 equations, and only 2 unknowns, a variable known as a Lagrange multiplier (L) can be added to minimise the estimation error, thus:

$$\begin{aligned} W_1 g(h_{11}) + W_2 g(h_{12}) + L &= g(h_{1p}) \\ W_1 g(h_{12}) + W_2 g(h_{22}) + L &= g(h_{2p}) \\ W_1 + W_2 &= 1 \end{aligned}$$

These equations can be rearranged in matrix form as:

$$\begin{vmatrix} g(h_{11}) & g(h_{12}) & 1 \\ g(h_{11}) & g(h_{12}) & 1 \\ 1 & 1 & 0 \end{vmatrix} \cdot \begin{vmatrix} W_1 \\ W_2 \\ L \end{vmatrix} = \begin{vmatrix} g(h_{1p}) \\ g(h_{2p}) \\ 1 \end{vmatrix}$$

or

$$|SV| \cdot |W| = |PSV|$$

The matrix equation is solved in the code by inverting SV to ISV, and multiplying ISV and PSV to give W. Both z_p and SE_p are then solved as follows:

$$\hat{z}_p = W_1 * z_1 + W_2 * z_2$$
$$SE_p = (W_1 * g(h_{1p}) + W_2 * g(h_{2p}) + L * 1)^{0.5}$$

"Krige2" subroutine

The Kriging subroutines are similar, with the matrices being sequentially increased from 1 to 5 stations. Continuing with the 2-station example, the steps are as follows:

1. Coordinates (x_i, y_i, z_i) of 2 stations are read from the Design sheet, and the grid points (x_p, y_p) are defined in a double nested Do Loop.
2. Separation distances (h) are coded.
3. Semivariances are coded by multiplying h by the SV slope (K), and the range (R) is defined for which the semivariance becomes constant at K*R.
4. Matrix SV is defined as an array.
5. Matrix SV is inverted to ISV.
6. Matrix ISV is multiplied by the semivariance matrix (PSV) to give weights and L.
7. z and SE are solved for the grid.

The SV slope and range are varied by reading the Spin button values from the "Run" worksheet.

Numerical Check

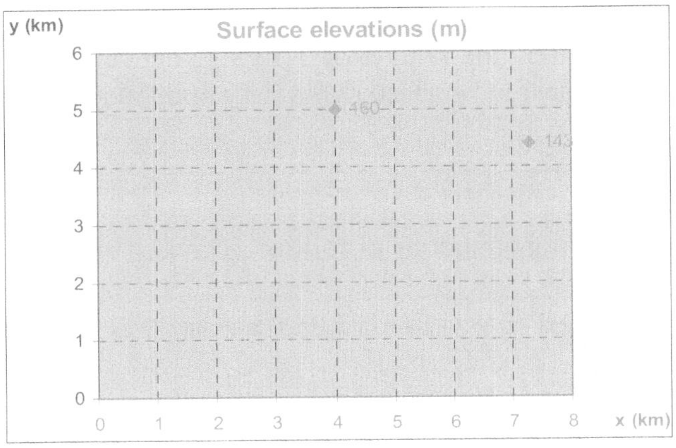

Figure 2.5.2: Plan for 2 stations

A numerical check on one grid point can be made with K = 4 m^2/km and R = 10 km as follows:

1. Coordinates (x_i, y_i, z_i) for 2 stations are 4, 5, 160 ; 7.3, 4.4, 143 (Figure 2.5.2)
   Use one grid point (x_p, y_p) as 0, 0.

2. Evaluate simultaneous equations:

$$W_1 g(h_{11}) + W_2\, g(h_{12}) + L = g(h_{1p})$$
$$W_1 g(h_{12}) + W_2 g(h_{22}) + L = g(h_{2p})$$
$$W_1 \quad + \quad W_2 \quad = 1$$

Since $h_{12} = 3.354$; $h_{1p} = 6.403$; $h_{2p} = 8.523$
Then $g(h_{12}) = 4 * 3.354 = 13.416$; $g(h_{1p}) = 25.612$; $g(h_{2p}) = 34.092$
The equations become:

$$0 + W_2 * 13.416 + L = 25.612$$
$$W_1 * 13.416 + 0 + L = 34.092$$
$$W_1 \quad + \quad W_2 = 1$$

Substituting variables we get:

$$W_1 = 0.816;\ W_2 = 0.184;\ L = 23.143$$

3. Thus z_p = 156.872 m and SE_p = 7.093 m as computed.

2.5.4 Universal Kriging for 3 stations

Theory

At least 3 data points are required for Universal Kriging, hence this is the simplest condition. The residuals of a linear trend surface (M) at any point may be given by:

$$M = a_1 x + a_2 y$$

This is incorporated into the Kriging equations as follows (Davis, 1986):

$$\begin{array}{llllll}
W_1 g(h_{11}) + W_2 g(h_{12}) & + W_3 g(h_{13}) + L & + a_1 x_1 & + a_2 y_1 & = & g(h_{1p}) \\
W_1 g(h_{12}) + W_2 g(h_{22}) & + W_3 g(h_{23}) + L & + a_1 x_2 & + a_2 y_2 & = & g(h_{2p}) \\
W_1 g(h_{13}) + W_2 g(h_{23}) & + W_3 g(h_{33}) + L & + a_1 x_3 & + a_2 y_3 & = & g(h_{3p}) \\
W_1 \quad + W_2 & + W_3 \quad + 0 & + 0 & + 0 & = & 1 \\
W_1 x_1 \quad + W_2 x_2 & + W_3 x_3 \quad + 0 & + 0 & + 0 & = & x_p \\
W_1 y_1 \quad + W_2 y_2 & + W_3 y_3 \quad + 0 & + 0 & + 0 & = & y_p
\end{array}$$

Thereby giving the matrices:

$$\begin{vmatrix}
g(h_{11}) & g(h_{12}) & g(h_{13}) & 1 & x_1 & y_1 \\
g(h_{12}) & g(h_{22}) & g(h_{23}) & 1 & x_2 & y_2 \\
g(h_{13}) & g(h_{23}) & g(h_{33}) & 1 & x_3 & y_3 \\
1 & 1 & 1 & 0 & 0 & 0 \\
x_1 & x_2 & x_3 & 0 & 0 & 0 \\
y_1 & y_2 & y_3 & 0 & 0 & 0
\end{vmatrix} \cdot \begin{vmatrix} W_1 \\ W_2 \\ W_3 \\ L \\ a_1 \\ a_2 \end{vmatrix} = \begin{vmatrix} g(h_{1p}) \\ g(h_{2p}) \\ g(h_{1p}) \\ 1 \\ x_p \\ y_p \end{vmatrix}$$

or

$$|SV| \cdot |W| = |PSV|$$

As with Ordinary Kriging, matrix SV is inverted and multiplied with PSV to give matrix W. The grid point solution is then given by:

$$zp = W_1 * z_1 + W_2 * z_2 + W_3 * z_3$$

$$SE_p = (W_1 * g(h_{1p}) + W_2 * g(h_{2p}) + W_3 * g(h_{3p}) + L * 1 + a_1 * x_p + a_2 * y_p)^{0.5}$$

Note that the SE is larger than for Ordinary Kriging, since it includes the uncertainty in the linear trend surface.

"Krige3U" subroutine

The Ordinary Kriging codes are held in the "Ordinary" module and the Universal Kriging are in the "Universal" module of this project. The Universal code is similar to the 7 steps for Ordinary Kriging, except that the additional trend components are added to the matrices as above. Again, the SV slope and range may be varied by the use of the Spin buttons on the Run Sheet.

Numerical Check

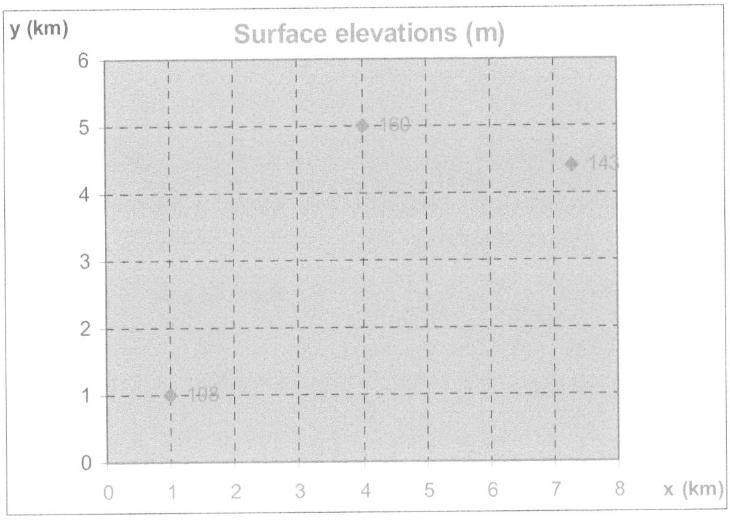

Figure 2.5.3: Plan for 3 stations

A numerical check on one grid point can be made with K = 4 m^2/km and R = 10 km:

1. Coordinates (x_i, y_i, z_i) for 3 stations are 4, 5, 160 ; 7.3, 4.4, 143 ; 1, 1, 198 (Figure 2.5.3). Use one grid point (x_p, y_p) as 4, 5 (which is chosen as equal to say x_1,y_1).
2. Evaluate the simultaneous equations using:

$$g(h_{12}) = 13.16 \; ; \; g(h_{13}) = 20 \; ; \; g(h_{23}) = 28.636$$
$$g(h_{1p}) = 0 \; ; \; g(h_{2p}) = 13.416 \; ; \; g(h_{3p}) = 20$$

$$\begin{array}{lllllll}
 & 13.416W_2 & +20W_3 & +L & +4a_1 & +5a_2 & = 0 \\
13.416W_1 & & +28.636W_3 & +L & +7.3a_1 & +4.4a_2 & = 13.416 \\
20W_1 & +7.159W_2 & & +L & +a_1 & +a_2 & = 20 \\
W_1 & +W_2 & +W_3 & & & & = 1 \\
4W_1 & +7.3W_2 & +W_3 & & & & = 4 \\
5W_1 & +4.4W_2 & +W_3 & & & & = 5
\end{array}$$

The last 3 equations may be used to solve for the weights only, and it is found:

$W_1 = 1; W_2 = 0; W_3 = 0$

3. Thus z_p = 160 m which is the true z_i value, and it follows that SE_p = 0 m.
The SE value may be proved by solving the first 3 equations to show $L = a_1 = a_2 = 0$.
It may also be verified in the computed result by using the Debug menu, and Run to Cursor command for a loop termination of (4, 5).

This example is the same as the computed result, and shows how the Universal Kriged surface is coincident with the 3 data points. Real physical surfaces are Kriged with many more data points, and the possible application of quadratic or cubic trend surfaces. It is then simpler to analyse the trend surface and residuals separately (as in the "Rainfall.xls" and "GWlevels.xls" projects given later), rather than include them in the same matrix.

2.6 BLOCK KRIGING

2.6.1 Summary

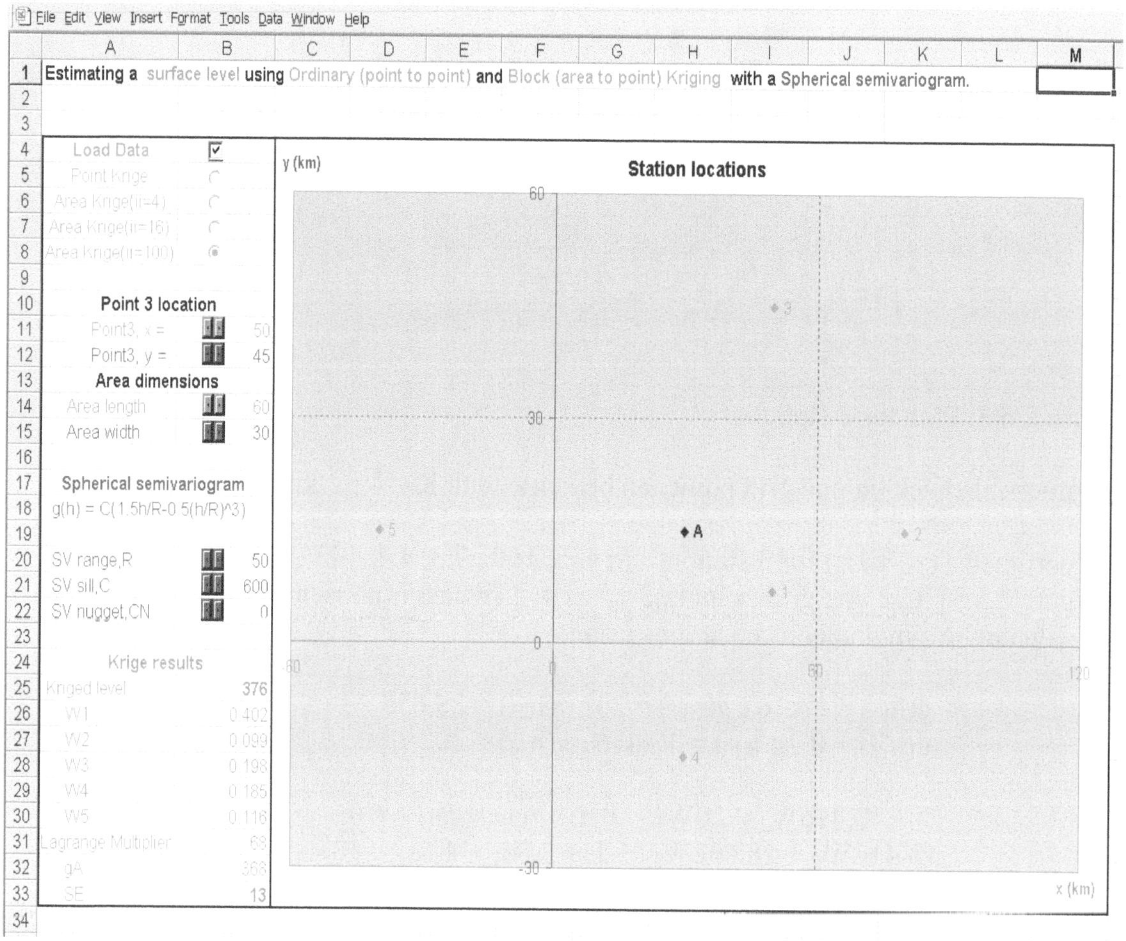

Figure 2.6.1: "Run" worksheet for the "BKrige.xls" code

Point Kriging includes both Ordinary and Universal Kriging, and is usually used to interpolate point data onto a grid. Block Kriging can be used for the same purpose, but the grid is no longer a regular set of points. It is a regular set of areas centered on the grid points. Block Kriging may also be used for one-dimensional and three-dimensional grids; it is however restricted herein to two-dimensional grids for simplicity.

The theory involves evaluating the semivariance functions between the areas to be Kriged and the data points (Isaaks and Srivasta, 1989). Since each block or area involves an averaging process of many points within the area, the standard error will be less than for Point Kriging. This result is demonstrated in the present project, which contrasts the results of Ordinary and Block Kriging for a single, variable area (centred at A) and 5 data points. The data used can be assumed to represent surface levels in metres on a kilometre based grid (Figure 2.6.1). This example is developed from an exercise given by Clark (personal communication).

2.6.2 Code operation

All data points are loaded in the "Design" worksheet by means of the "Load data" Check box, and the Option buttons are used to select the desired type of Kriging. Block Kriging involves the double and quadruple integration of the semivariances (g) for the area, and this is approximated by the discretisation of many points (ii) in the area. The successive approximation of using ii = 4, 16 and 100 points may be examined by selection of the Option buttons. It is usually found that 16 points are adequate, but this will depend on the data and level of required accuracy.

Since the surface level is considered continuous between each data point, it is spatially correlated over a short distance known as the range (R). This correlation is measured by the semivariogram (SV) of a regionalised variable and is assumed to be of the spherical type (Clark, 2000), as illustrated below. (There are three common models used for a SV, namely linear, spherical and exponential. The simpler use of a polynomial SV is also demonstrated later in the "GWLevels.xls" project). The hypothetical SV reaches a constant value (C+CN) for an average distance interval between points of h = R; it may also have a "nugget" effect (a non-zero intercept, CN) such that it is discontinuous at h = 0. The 3 variables, R, C and CN may be varied by means of the Spin button controls.

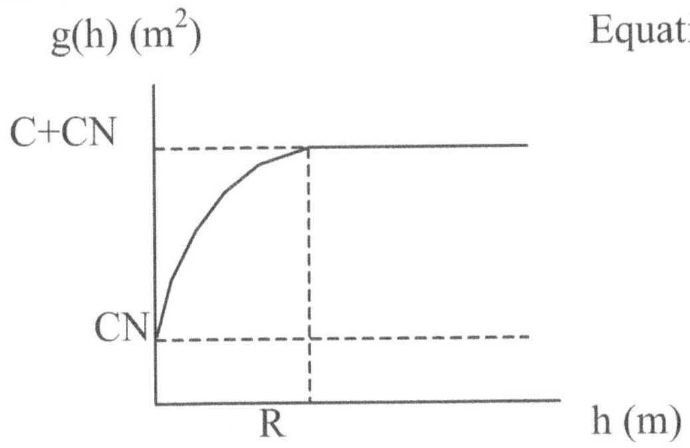

Equation:

$$g(h) = 0 \quad \text{for } h = 0$$
$$g(h) = CN + C(1.5h/R - 0.5(h/R)^3) \quad \text{for } R>h>0$$
$$g(h) = C+CN \quad \text{for } h>R$$

Figure 2.6.2: A spherical semivariogram

The Option buttons operate the Kriging code according to the SV variables, and the resultant Kriged surface level, data point weights and standard error are written to the "Run" worksheet. The influence of moving a data point (say Point 3), changing the Kriged area and the SV variables may be examined by operating the associated Spin buttons. Two examples are given in Table 2.6.1 that illustrate the reduction in SE with Block Kriging.

Kriging							Point	ii = 100
Point3, x	Point3, y	Length	Width	SV Range	SV sill	SV nugget	Level ±SE	Level ±SE
40	45	60	30	100	700	100	377 ± 20	375 ± 11
40	45	40	20	70	500	0	353 ± 19	355 ± 14

Table 2.6.1: Example results for Point (ii = 1) and Block Kriging (ii=100)

2.6.3 Theory

The Block Kriging matrix formulation can be derived from the Ordinary Kriging matrix formulation as given in the "PointKrige.xls" theory previously (Isaaks and Srivastava, 1989). For the Ordinary Kriging of 2 stations, we had:

$$\begin{vmatrix} g(h_{11}) & g(h_{12}) & 1 \\ g(h_{11}) & g(h_{12}) & 1 \\ 1 & 1 & 0 \end{vmatrix} \cdot \begin{vmatrix} W_1 \\ W_2 \\ L \end{vmatrix} = \begin{vmatrix} g(h_{1p}) \\ g(h_{2p}) \\ 1 \end{vmatrix}$$

or $$|SV| \;.\; |W| = |PSV|$$

The matrix equation is solved by inverting SV to ISV, and multiplying ISV and PSV to give W. Both z_p and SE_p are then solved as follows:

$$z_p = W_1 * z_1 + W_2 * z_2$$
$$SE_p = (W_1 * g(h_{1p}) + W_2 * g(h_{2p}) + L * 1)^{0.5}$$

Since the location of the Kriged point (p) or area (a) is independent of matrix $|SV|$, the latter remains unaltered for Block Kriging. The $|PSV|$ matrix is however changed for Block Kriging as the semivariance terms are now between each data point and the Kriged area ($g(h_{1a})$ and $g(h_{2a})$ say). The resultant semivariance is thus the integration of the average semivariance between each data point and the area, thus $g(h_{1a})$ becomes:

$$g(h_{1a}) = (1/a)\ {}_x\!\int {}_y\!\int g(h(p_1 - u_{x,y}))\,dxdy \qquad \text{where } p_1 = x_1, y_1;\ a = xy$$

This double integral may be discretised by dividing the area, a, into N subareas $A_1, \ldots. A_N$. and representing each subarea, A_k, by a central point u_k (Kitanidis, 1997), thus:

$$g(h_{1a}) = (1/a)\ {}_{k=1}\sum^{N} g(h(p_1 - u_k))A_k$$

And since ${}_{k=1}\sum^{N} A_k = a$

$$g(h_{1a}) = (1/N)\ {}_{k=1}\sum^{N} g(h(p_1 - u_k))$$

Thus the Block Kriging area-point semivariance becomes the average semivariance between a data point and all points within the area. Inevitably, the accuracy will improve as $N \rightarrow \infty$, and this may be examined in the exercise code for N = ii = 4, 16 and 100, or by changing ii in the "Functions" module. The area-point semivariance is illustrated in Figure 2.6.3 for N = 4 subareas.

Once the area-point semivariances are substituted in the $|PSV|$ matrix, the solution of z_A for the centre of the area (x_A,y_A) follows that of Ordinary Kriging.

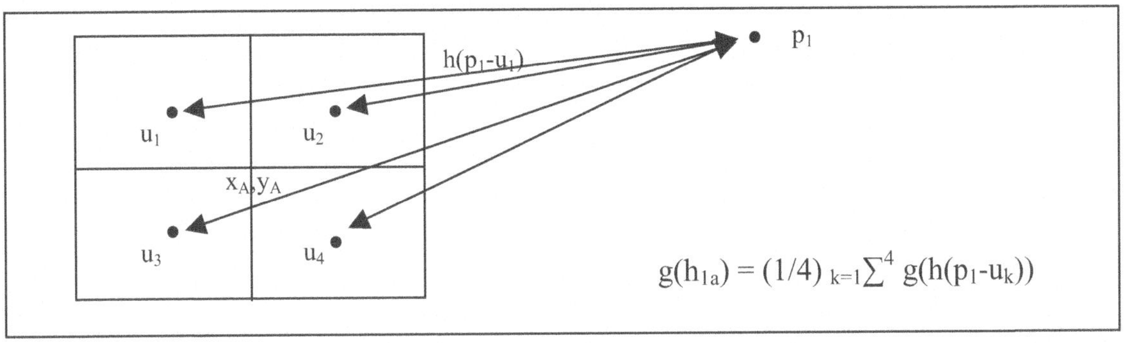

Figure 2.6.3: Area-point semivariance for N = 4

The standard error (SE) for the Block Kriging estimate differs slightly from that for Ordinary Kriging. In addition to the changed $|PSV|$ values, the SE is reduced by the quadruple integral of all point semivariances within the local area:

$$gA = (1/a^2) \int_{x1} \int_{y1} \int_{x} \int_{y} g(h(u_{x1,y1} - u_{x,y}))\, dxdy\, dx_1dy_1$$

As with the above double integral for $g(h_{1a})$, the quadruple integral (gA) may be discretised as the average of all pairs of locations within the area, thus:

$$gA = (1/a^2) \sum_{i=1}^{N} \sum_{j=1}^{N} g(h(u_i\text{-}u_j))\ A_iA_j$$

And since $\sum_{i=1}^{N} A_i = \sum_{j=1}^{N} A_j = a$

$$gA = (1/N^2) \sum_{i=1}^{N} g(h(u_i\text{-}u_j))$$

Thus the SE is reduced by the mean semivariance of all data points within a local area:

$$SE_A = (W_1 * g(h_{1a}) + W_2 * g(h_{2a}) + L * 1 - gA)^{0.5}$$

As with the $|PSV|$ values, gA is approximated using N = 4, 16 and 100 values in the Block Kriging exercise code. This area-area semivariance is illustrated in Figure 2.6.4 for N = 4 subareas.

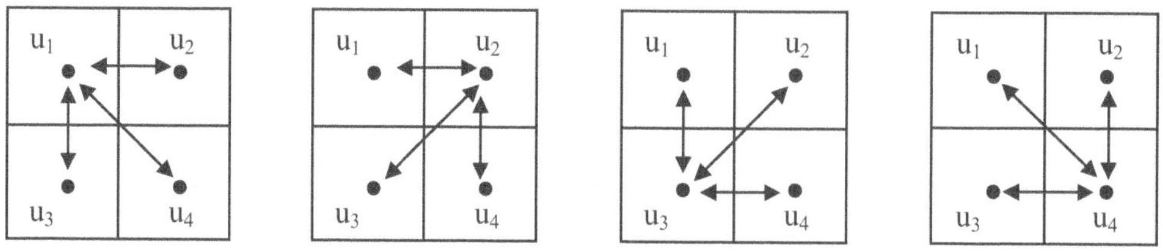

Figure 2.6.4: Area – area semivariance for N = 4

2.6.4 Modules

"Krige"

The "Krige" module consists of 2 main subroutines, namely "Krige5P" and "Krige5A". "Krige5P" is similar to that for Point Kriging with 5 stations, and is left herein for comparison. It can however be simply verified, by substituting ii = 1 in the "Krige5A" code. The Point Kriging code is thus equivalent to Block Kriging when one point only is used for the Kriged area. In the following codes for "GWLevels.xls" and "UEden349.xls", the single Block Kriging code is used throughout.

The "Krige5A" subroutine has been expanded to 8 steps, after the SV variables (R, C, CN), grid-area central point variables (x_A, y_A) and grid-area dimensions (ml, mb) have been read. These steps are:

1. Coordinates (x_i, y_i, z_i) of the 5 stations are read from the "Design" worksheet.
2. Separation distances (h) are coded. Note that the coding is now in the form of a 6 * 6 array (or matrix, H), and this is a precursor for looping the code for many grid-area points. The SV boundary condition is also included in this step; it simply states that if the distance between points exceeds R, then it is made equal to R.
3. Semivariances are coded by applying the SV formula. Again, the coding is formed as a 6 * 6 array (G), and all arrays are declared at the top of the module. The area-point semivariances are also included as an array (PSV) in this step; they are calculated as a function procedure (described below).
4. Matrix SV is defined as an array. Note that when CN is finite, G(i,i) evaluates to zero. The discontinuity for the spherical SV at h = 0 is thus maintained by declaring SV(i,i) =0.
5. Matrix SV is inverted to ISV and matrix ISV is multiplied by matrix PSV to give the weights and L.
6. z_A and SE are solved after the area-area (gA) semivariance is evaluated using a function procedure.
7. Kriged results are written to the "Run" Worksheet.
8. Arrays may be written to the "Design" Worksheet for numerical checking.

"Functions"

Two functions are called by the "Krige" module, and these are coded as function procedures "PSVF" and "gAF". The procedures have the advantage of reducing code for the "Krige" subroutine, and accepting the necessary variables. "PSVF" operates with 9 relevant variables, namely x_A, y_A, ml, mb, x, y, R, C and CN. Variables x_A and y_A are the coordinates of the area centre, and x and y are those of the consecutive data points. Variables ml and mb are the area dimensions, and R, C and CN define the SV. The function operates in 4 steps, as follows:

1. Variable ii is first selected using a multiple If condition for the Option Button chosen in the "Run" worksheet. Matrices XAX and XAY represent the area grid points; they have been declared as open matrices at the top of the module. They are now redimensioned according to the ii selection. Matrices HA and PSVA represent the distance between points and their associated semivariance respectively; they are similarly redimensioned.
2. The XAX grid points are generated as central points for each subarea using a double loop
3. The XAY grid points are generated as central points for each subarea using a double loop
4. The distances, HA, between each area point and the data points are calculated, and the associated semivariances are summed in a single loop. The final statement averages the sum as the variable PSVF. Note the range boundary condition for HA = R and the nugget effect for HA = 0. (The latter is necessary for Point Kriging using ii = 1, since PSV(i,i) = 0 and gA(i,i) = 0.)

The gAF function follows the same steps as for PSVF, except that step 4 is now a double loop for the area points related to themselves. Both functions may be numerically checked using the Test subroutines. The latter conclude this module.

Numerical check

The check is applied for N = 4; Point3 = 40, 45 km ; ml = 40 km, mb = 20 km; R = 70 km, C = 500 m^2 and CN = 0 (Figure 2.6.5). A direct method for checking complex matrices with computer aid is to use the Debug facility in the VBE. The "Krige" module is activated, and the cursor is clicked at the End Sub line of the "Krige5A" subroutine. With the commands Debug.Run to Cursor, the code is activated and all variables may be examined and checked by substitution. The 6 * 6 matrix formulation was thus read as:

$$
\begin{aligned}
0 + 310.8W_2 + 376.8W_3 + 299.4W_4 + 500.0W_5 + L &= 300.7 \\
310.8W_1 + 0 + 444.6W_3 + 480.2W_4 + 500.0W_5 + L &= 473.6 \\
376.8W_1 + 444.6W_2 + 0 + 487.7W_4 + 500.0W_5 + L &= 387.5 \\
299.4W_1 + 480.2W_2 + 487.7W_3 + 0 + 500.0W_5 + L &= 287.2 \\
500.0W_1 + 500.0W_2 + 500.0W_3 + 500.0W_4 + 0 + L &= 473.6 \\
W_1 + W_2 + W_3 + W_4 + W_5 + 0 &= 1
\end{aligned}
$$

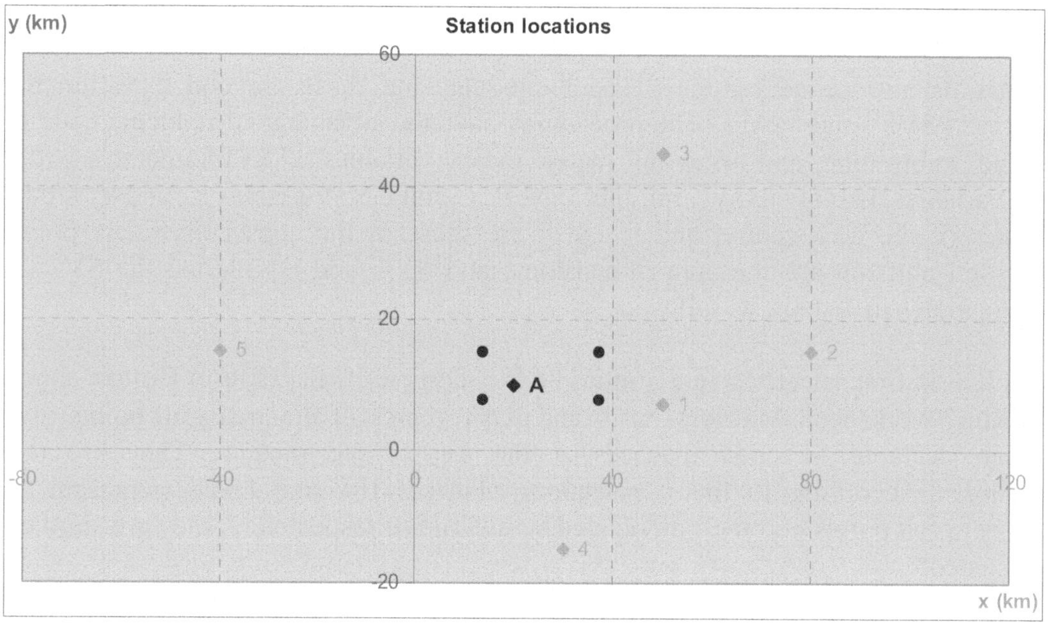

Figure 2.6.5: Check of block Kriging using N = 4

Example checks are as follows:

$H(1,2) = ((80\text{-}50)^2 + (15\text{-}7)^2)^{0.5} = 31.1$; $G(1,2) = 500(1.5(31.1/70) - 0.5(31.1/70)^3) = 311.2$

$H(1,A) = 40.8, 21.5, 40.1, 20.1$; $G(1,A) = 387.6 + 223.1 + 382.6 + 209.4$
$PSV(1,A) = G(1,A)/4 = 300.7$

$HA = 4(0, 20, 10, 22.36)$; $GA = 4(0 + 208.5 + 106.4 + 231.4)$
$gA = GA/16 = 136.6$

The weight solutions can now be used to verify the matrices. The variables zA and SE are calculated as:

$zA = 0.267*400 - 0.013*380 + 0.227*450 + 0.389*280 + 0.129*320 = 354.3$ m

$SE^2 = 0.267*300.7 - 0.013*473.6 + 0.227*387.5 + 0.389*287.2 + 0.129*473.6 + 38 - 137$
$SE = 15.4$ m

Note that a weight can sometimes be small and negative. This is a consequence of the matrix calculation, especially when the coefficients are limited to C by the range condition.

2.6.5 Controls

Option buttons

The influence of the number of points (N or ii) used in an area for Block Kriging is readily assessed using the Option Button controls. Continuing the above example (Point3 = 40, 45 km ; ml = 40 km, mb = 20 km; R = 70 km, C = 500 km^2 and CN = 0), the following results are obtained:

Krige method	Point	Area ii = 4	Area ii = 16	Area ii = 100
zA	353.1	354.6	354.9	355.0
SE	19.1	15.4	14.7	14.5

Table 2.6.2: Influence of ii in block Kriging

The results converge rapidly, and ii = 16 gives a suitable result for integer accuracy. Both Point and Block Kriging give similar levels, but the standard error is always less for the Block method.

"Point3" Spin buttons

The coordinates of Point 3 can be varied from 0 km to 100 km with the Spin button controls. Increasing the coordinates decrease the associated weights, and vice-versa.

The above example illustrates how Point 1 (which is closer to Point A) shelters Point 2, since they are of similar direction. The effect can be better demonstrated by spinning Point 3 to 40, 10. The point is now closest to Point A, taking a weight of 0.623 for ii = 100, whilst W_1 and W_2 are negligible. Note that if Point 3 = Point A, its weight for Point Kriging is unity, whereas the weight decreases from unity for Block Kriging since area averages are now involved.

"Area" Spin buttons

The area coordinates may be varied from 5 km to 100 km with the Spin button controls. A zero value cannot be used in the code as it produces a chart plotting error.

For the largest area (100 km * 100 km), the SE is least at 9 km (ii = 100) since gA is a maximum. For the smallest area (5 km * 5 km), the SE is largest at 21 km (ii = 100), since gA is minimum. Note that the Point Kriging solution is almost the same for the smallest area, since the data points are furthest separated, and the area is approaching a point.

"Range" Spin button

The range coordinate may be varied from 5 km to 200 km with the Spin button control. The value of zero cannot be used as it gives a "divide by zero" error in the code for the h/R term.

Since the furthest distance between data points (2 and 5) is 120 km, it is expected that decreasing R from 200 km to 120 km has little influence on zA (zA = 355 m for ii = 100). As R is decreased below 120 km, the SV matrix becomes influenced by maximum C values, zA increases and the weights approach equality. By R = 10 km, the SV matrix has become symmetric (Kitanidis, 1997) and zA has increased to its mean value of 366 m. For R = 5 km, the solution is almost that of the pure nugget effect (C = 0, CN = constant, demonstrated below), since all points (except those in the gA function) are unaffected by the semivariance.

"Sill" Spin button

The SV sill value (C) may be varied from 0 m^2 to 1000 m^2 with the Spin button control. Values of C = 0 and CN = 0 cannot be used together, as they give a zero SV matrix which cannot be inverted. The C and CN control codes (in the Sheet1 (Run) object) are therefore protected by a multiple If statement. (Reducing C does not affect the weight values and zA, but the SE is reduced since the PSV and gA functions are reduced.)

"Nugget" Spin button

The SV nugget value (CN) may be varied from 0 m^2 to 1000 m^2 with the Spin button control. In general, increasing the nugget value increases all variables since all matrix coefficients are increased.

If C is zero, the results display a "pure nugget" effect for which the data is unaffected by the semivariance and behaves as an ordinary random variable. This is illustrated in the following example for 2 stations:

$$W_1 g(h_{11}) + W_2 g(h_{12}) + L = g(h_{1a})$$
$$W_1 g(h_{12}) + W_2 g(h_{22}) + L = g(h_{2a})$$
$$W_1 \quad + \quad W_2 = 1$$

Thus

$$0 \quad + 500W_2 + L = 500$$
$$500W_1 + 0 \quad + L = 500$$
$$W_1 \quad + \quad W_2 = 1$$

Whence $W_1 = W_2 = 0.5 = 1/\text{no.stations}$ And $L = 250 = \text{CN/no. stations}$

For this project code, using C = 0 and CN = 500 we have:

$$W = 1/5 = 0.2,\ zA = \sum z/5 = 366\text{ m}$$
$$SE_A = (W_1 * g(h_{1a}) + W_2 * g(h_{2a}) + L * 1 - gA)^{0.5}$$
$$= (5(0.2 * 500) + 500/5 - 500)^{0.5} = \underline{10\text{m}}$$

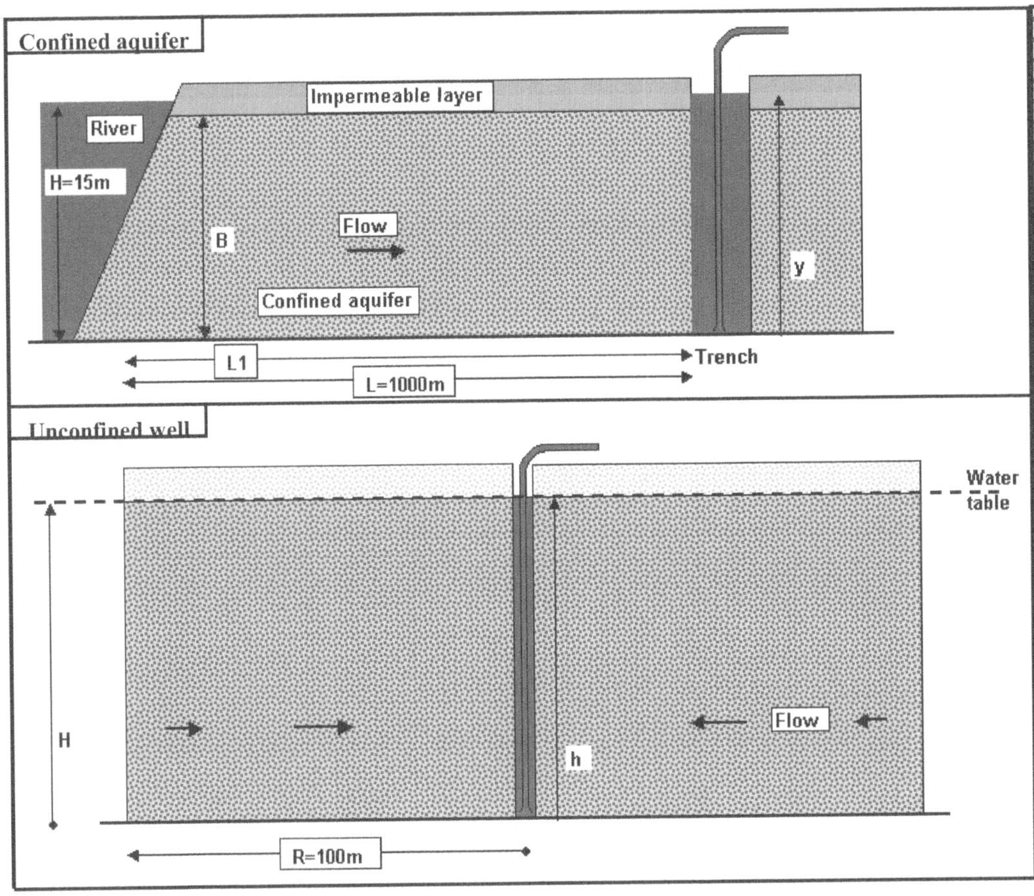

Groundwater analysis

3

3.1 GROUNDWATER LEVELS BY POINT AND BLOCK KRIGING

3.1.1 Summary

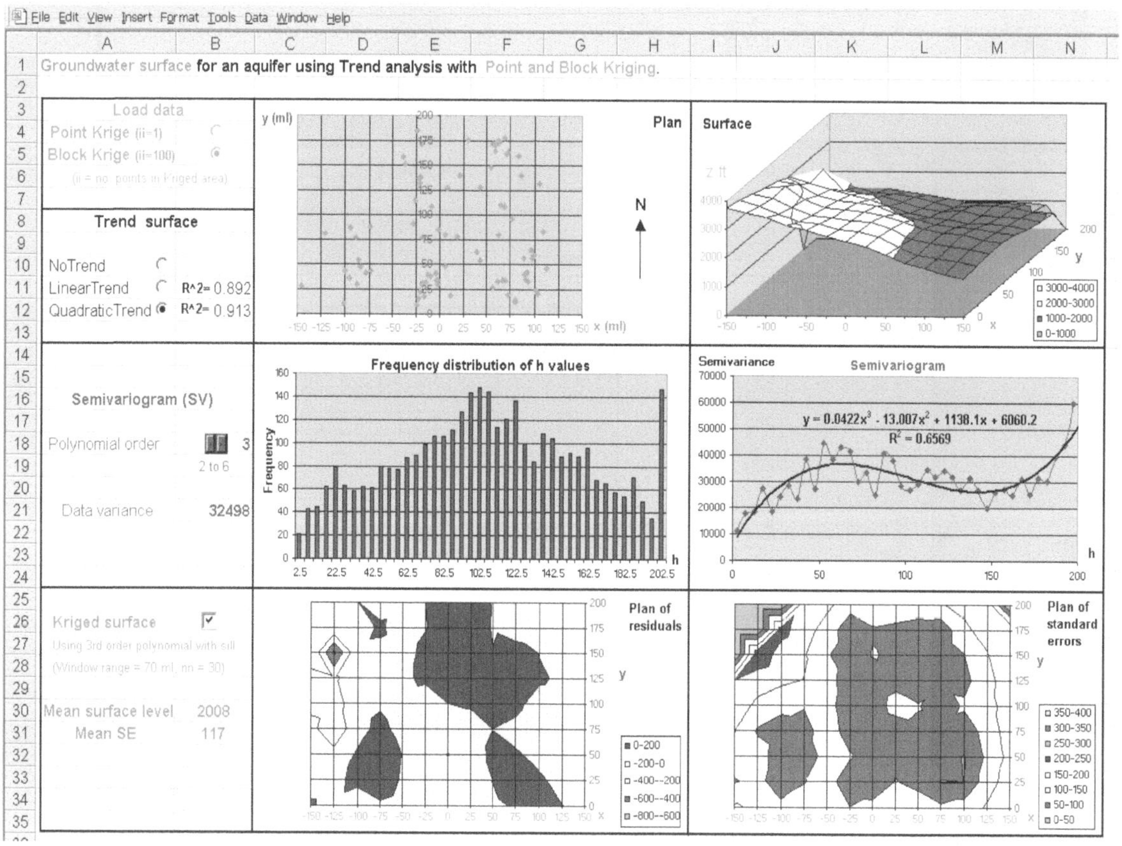

Figure 3.1.1: The "Run" worksheet for the "GWL.xls" project

This project involves the interpolation of groundwater levels with their associated standard error from point samples, for an aquifer in North Texas, USA. (The data was obtained with permission from Clark et al, 2000.) It illustrates the derivation and use of a semivariogram in Kriging analysis. The Kriging codes in Chapter 2 used a linear and spherical semivariogram by assumption. Herein, the theory is developed in more detail and the concepts of adding linear and quadratic trends to the data are explained. The residuals are then Kriged using either the Point or Block method.

The project sequence is demonstrated in 6 charts (Figure 3.1.1). The "Plan" chart shows the horizontal coordinates (x, y in miles) of the original point data. The "Surface" chart plots the 3 dimensional data with or without a computed trend, and with or without Kriging. The "Frequency distribution" chart and "Semivariogram" chart display the results of the Kriging analysis, as described below. The final result is the estimation of the Kriged data residuals from the trend and the associated standard error. These are shown in the lower charts. An estimate for the mean groundwater, surface level with standard error is finally computed.

3.1.2 Code operation

There are 4 separate control areas in the control panel on the left hand side of the "Run" worksheet. These contain control choices of Option buttons, Spin buttons or Check boxes. The controls are operated in sequence from top to bottom, and any non-sequential entry triggers a Message Box warning that is generated by the control subroutines in the Sheet1 (Run) object. The control areas are now described in sequence.

"Load data" area

The 3 dimensional data (x, y and z, where z is the groundwater level) are written to the "Data" worksheet whenever the "Point Krige" or "Area Krige" Option button is activated. This data may be overwritten with alternative data in the subroutine "LoadData" of the "Data" module. The level locations (x, y) are also plotted on the "Run" worksheet.

"Trend surface" area

When Option Button "NoTrend" is activated, 3 subroutines are run sequentially. The first subroutine, "NoTrend", maps the data onto a 300 mile by 200 mile grid at 25 mile orthogonal intervals. The data is interpolated according to the elevation of its nearest neighbour, written to the "Design" worksheet, and plotted as an Excel surface chart in the "Run" worksheet. Note the northeast-southwest trend of increasing levels.

The second subroutine, "HistNoTrend", computes an array of the independent x, y distances (h) between all the n points (totalling n(n-1)/2 distances) and the corresponding square of the elevation differences (d). The third subroutine, "SV", computes the actual histogram of the semivariances (explained below), and all arrays are written to the "SV" worksheet. The histogram and the semivariance (g) curve are plotted in the "Run" worksheet, and a polynomial trendline is fitted to the semivariogram. The polynomial order may be adjusted from 2 to 6 using Spin Button 1. Note the continuous increase of the semivariogram, indicating that the data is non-stationary and trend analysis must be undertaken.

Option Button "LinearTrend" runs 3 corresponding subroutines in the "Trends" module for a hypothesised linear trend. These are "LinearTrend", "HistLTrend" and "SV" (as for "NoTrend"). The "NoTrend" data is replaced in the "Design" worksheet with a linear trended grid, and the SV data is replaced in the "SV" worksheet. The 3 graphs of surface, histogram and semivariance are now replaced in the "Run" worksheet. Note that the largest abcisssa for the histogram sums all readings greater than the value, and is a simple means of truncating the graph. The semivariance curve, computed from the detrended data, is now typical of a stationary regionalised variable. It increases for an h range of about 70 miles, after which it is approximately constant (and approximately equal to the data variance) until the few final readings provide spurious values. The simplest polynomial order of 3 is chosen for Kriging.

Option Button "QuadraticTrend" may be activated to compute a quadratic trend analysis which is programmed in the "Quadratic" module. A "goodness of fit" indicator (R^2) is computed for both linear ($R^2 = 0.892$) and quadratic trend ($R^2 = 0.913$), showing the latter to be a better fit. Again the semivariance curve shows stationarity and the simplest polynomial fit is of third order.

"Semivariogram" area

As described above, the Spin button selects the polynomial, and the data variance is written to this area.

"Kriged surface" area

Kriging is effected for the linear and quadratic trends using the "Krige surface" Check box to operate the "KrigeA" subroutine in the "Krige" module. The mapped data for residuals and standard error are written to the "Design" worksheet, and plotted in plan on the "Run" worksheet. The Kriged surface is computed using the subroutine "KrigedSurface" in the "Data" module, and plotted on the surface chart in the "Run" worksheet.

Summary of results

Point Krige (ii =1)	No Trend	Linear Trend	Quadratic Trend	**Area Krige (ii = 100)**	No Trend	Linear Trend	Quadratic Trend
Goodness of Fit (R^2)		0.892	0.913			0.892	0.913
Data Variance (VAR)	373737	40359	32498		373737	40359	32498
Mean surface level (MSL)	2002	2044	2009		2002	2043	2008
Mean standard error (MSE)	66	187	177		66	127	117

Table 3.1.1: Summary of "GWL.xls" results

If the data were considered as totally random and normally distributed, the mean surface level (MSL) with standard error is 2002 ± 66 feet. However, the data is neither random or normal and there exists a significant and similar linear or quadratic trend. These trends are computed and removed from the data before Kriging.

Residual Point Kriging maps this data onto a grid of points, and the resulting MSL and mean standard error (MSE) is averaged (for both trends) at 2027 ± 182 feet. Residual Block Kriging maps this data onto a grid of equal areas (25 square miles), and the resulting MSL and MSE are averaged (for both trends) at 2026 ± 122 feet. The MSL for Point and Block Kriging are similar since the grid area is much less than the data area ($25\ ml^2 \ll 30000\ ml^2$). The MSE for Block Kriging is however, less than that for Point Kriging due to the averaging of data within a grid area.

3.1.3 Modules

All control operations for this code are stored as Private subroutines in the Sheet1 (Run) object. The remaining subroutines are stored in 4 modules, namely "Data", "Functions", "Krige" and "Trends". The "Functions" module is detailed in the "BlockKrige.xls" project, and is therefore omitted herein.

"Trends"

The nearest neigbour elevation is computed in the "NoTrend" subroutine using an inner loop for each grid point, (x_p, y_p), and calculating the distance (h) of all neighbours (x_i, y_i), where:

$$h = ((x_p - x_i)^2 + (y_p - y_i)^2)^{0.5}$$

The h, x, y, z array is written to the "Data" worksheet (columns D:G), and sorted in ascending order to give z of the nearest neigbour in the first data row. As with the "PointKrige.xls" project, the grid is mapped using 2 nested outer loops for the Easting and Northings.

A linear trend surface is an equation of the type:

$$z = b_0 + b_1 x + b_2 y$$

It is calculated so that the squared deviations from the trend are minimised. Since the equation requires 3 unknown coefficients, 3 normal equations are required (Davis, 1986):

$$\begin{aligned} \sum z &= b_0 n + b_1 \sum x + b_2 \sum y \\ \sum xz &= b_0 \sum x + b_1 \sum x^2 + b_2 \sum xy \\ \sum yz &= b_0 \sum y + b_1 \sum xy + b_2 \sum y^2 \end{aligned}$$

The equations may be rewritten in matrix form as :

$$\begin{vmatrix} n & \sum x & \sum y \\ \sum x & \sum x^2 & \sum xy \\ \sum y & \sum xy & \sum y^2 \end{vmatrix} \cdot \begin{vmatrix} b_0 \\ b_1 \\ b_2 \end{vmatrix} = \begin{vmatrix} \sum z \\ \sum xz \\ \sum yz \end{vmatrix}$$

or matrices $A \cdot B = C$

Subroutine "LinearTrend" calculates the summation A and C matrices by looping for each component, then inverting A and multiplying by C to find B. The "goodness of fit" coefficient, R^2 is found by:

$$R^2 = \frac{\text{sum of squared residuals from the trend}}{\text{sum of squared residuals from the data}} = \frac{{}_n\sum (z_{est} - z_{bar})^2}{{}_n\sum (z_i - z_{bar})^2}$$

where z_{est} is the trended estimate and z_{bar} is the mean value of the z_i data. R is called the coefficient of multiple correlation, and varies like R^2 from 0 to 1.

Subroutine "LinearTrend" also writes the detrended values (z_{est} - z_i) to the "SV" worksheet, and uses the double nested loop to write grid values to the "Design" worksheet.

Subroutine "QuadTrend" operates as for the "LinearTrend" code, except the A matrix is now 6 * 6, and the other 2 matrices are 1 * 6 since the quadratic trend equation has 6 unknown coefficients:

$$z = b_0 + b_1x + b_2y + b_3x^2 + b_4y^2 + b_5xy$$

The matrix equation becomes:

$$\begin{vmatrix} n & \sum x & \sum y & \sum x^2 & \sum y^2 & \sum xy \\ \sum x & \sum x^2 & \sum xy & \sum x^3 & \sum xy^2 & \sum x^2y \\ \sum y & \sum xy & \sum y^2 & \sum x^2y & \sum y^3 & \sum xy^2 \\ \sum x^2 & \sum x^3 & \sum x^2y & \sum x^4 & \sum x^2y^2 & \sum x^3y \\ \sum y^2 & \sum xy^2 & \sum y^3 & \sum x^2y^2 & \sum y^4 & \sum xy^3 \\ \sum xy & \sum x^2y & \sum xy^2 & \sum x^3y & \sum xy^3 & \sum x^2y^2 \end{vmatrix} \cdot \begin{vmatrix} b_0 \\ b_1 \\ b_2 \\ b_3 \\ b_4 \\ b_5 \end{vmatrix} = \begin{vmatrix} \sum z \\ \sum xz \\ \sum yz \\ \sum x^2z \\ \sum y^2z \\ \sum xyz \end{vmatrix}$$

The code is thus extended from the 3*3 matrix form of LinearTrend.

"Data"

The "HistTrend" subroutine is computed on the "SV" worksheet, and is used to prepare data for the histogram chart on the "Run" worksheet. The independent distances between each data point (h) and associated square of their elevation difference (the elevation variance, d) are computed as:

$$h = ((x_j - x_i)^2 + (y_j - y_i)^2)^{0.5}$$
$$d = (z_j - z_i)^2$$

where the subscripts i and j denote different data points. The total of n(n-1)/2 values (3570 for n = 85) are generated by two outer nested loops, and a counter (k) is used to write the data sequentially to worksheet "SV". In the MS Excel 2000 code used herein, the n(n-1)/2 value denotes a limit of n = 362 data points which can be used for alternative data, before the spreadsheet row limit of 65,536 is exceeded. This n value is used to limit the copying and counting of columns throughout the code.

The "HistTrend" subroutine is common for all trends since the detrended elevation differences (z_{est} - z_i) are written to the "SV" worksheet by subroutines "LinearTrend" and "QuadTrend".

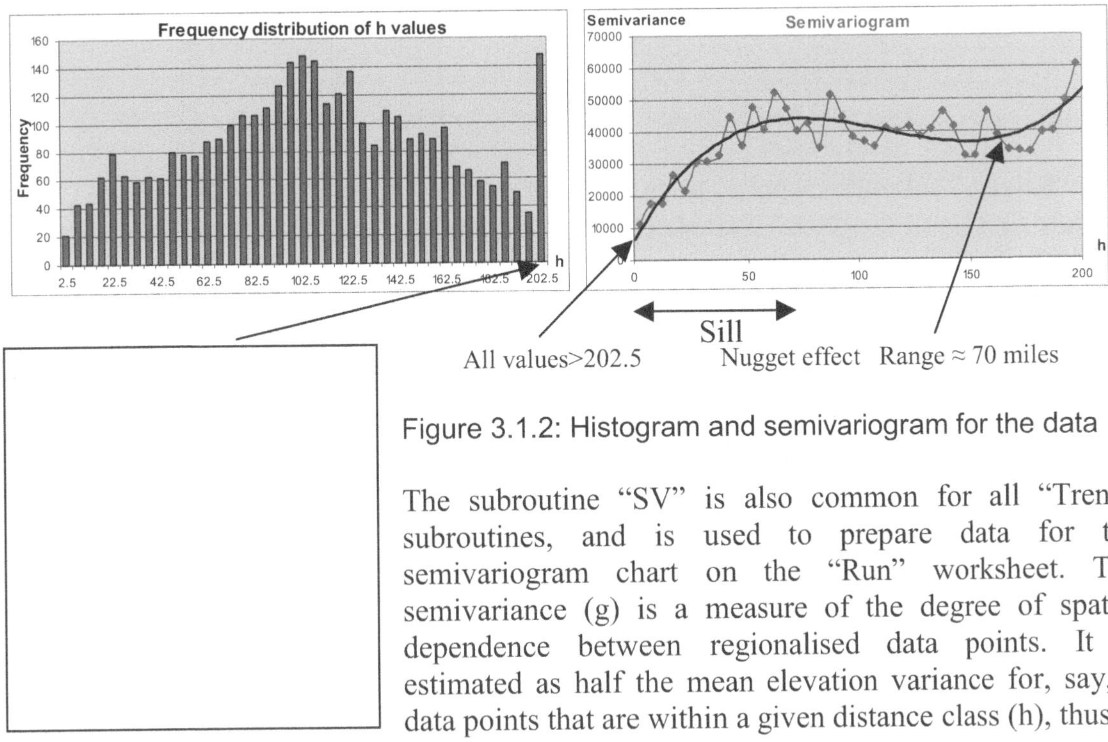

Figure 3.1.2: Histogram and semivariogram for the data

The subroutine "SV" is also common for all "Trend" subroutines, and is used to prepare data for the semivariogram chart on the "Run" worksheet. The semivariance (g) is a measure of the degree of spatial dependence between regionalised data points. It is estimated as half the mean elevation variance for, say, k data points that are within a given distance class (h), thus:

$$g(h) = {}_k\sum (z_i - z_{i+h})^2 / 2k$$

where variable z_i is at a position i and z_{i+h} is at a position (i + h) in a random field. The distance classes are first defined in the subroutine and written to "SV" as a column of histogram abcissae values (bins), using code from the worksheet [Edit, Fill, Series] command. (When using alternative data, these values may be changed manually, and it must be ensured that all bins are occupied.) The elevation variance is summed within each bin (Sd), and the number of readings counted (i). The subroutine loops through each bin value, and writes both bin values and semivariance to worksheet "SV". Note that the subroutine uses 2 pieces of code to speed execution. Firstly, the h and d data are sorted in ascending order so that the lowest bin values are quickly accessed, and then the accessed bin data is deleted after each bin computation.

The histogram and semivariogram for this data are shown in Figure 3.1.2. The histogram is truncated at the highest value by counting all the remaining points. The semivariogram increases for a certain range (about 70 miles), after which it remains approximately constant (at the so-called Sill) until the few remaining values cause oscillation. The presence of the sill indicates stationarity of the function ($\partial z/\partial x = \partial z/\partial y = 0$). The semivariogram does not pass through the origin (the nugget effect), and this indicates that the regionalised variable is highly variable over distances less than the bin interval.

Various linear, circular and exponential functions have been proposed to describe the semi-variogram increase (Davis, 1986). A simple polynomial fit is used herein, and the nugget effect is modelled by letting g(h = 0) = 0 in the gA and PSV functions of the "Functions" module.

"Krige"

The "KrigeA" subroutine applies to either Point or Block Kriging according to the Option button chosen on the "Run" worksheet. For Point Kriging, the "Functions" module uses a multiple If condition to set the gA and PSV functions to one point. For Block Kriging, the condition sets the number of points to say 100.

Since 85 data points are used for this exercise, the code is generalised to matrix form for Kriging. It is found in MS Office 2000 that matrices greater than about 60 * 60 become unstable, thus the code uses a moving neighbourhood range (called a "Window range") to restrict a matrix size. Furthermore, a restriction on the number of nearest neighbours (nn) used for Kriging may also be made to speed execution.

The generalised code is essentially a grid application of the "BKrige.xls" code described previously, and may be summarised in 12 steps:

1. The SV variables are set according to the linear or quadratic trend selected. Since a polynomial function is used, the nugget effect is included by ensuring the matrix, SV, has a zero diagonal.
2. The data is counted (n points) and grid points, xA and yA, are set by means of double loops surrounding the code. The gA function is also evaluated in this step, as it is unique for each grid area.
3. The distances between the grid and data points (hh) are calculated, and written to column 4 in the "Data" worksheet.
4. Data points x, y and z (residual) are written to the "Data" worksheet, and hh, x, y and z (residual) are sorted in ascending order for later selection in Kriging.
5. nn points are selected for a Window range of say WR = 70 miles, and nn is further restricted to say 30 points.
6. The dynamic arrays for SV, PSV, H (distance between points) and XXA (x, y, z coordinates of data points) are redimensioned.
7. Matrix XXA is read, and the PSV matrix is formulated from the function subroutine.
8. The semivariance matrix is solved in a double loop, and the range boundary condition applied for H > R.
9. Matrix SV is inverted, and matrix W is solved by multiplication.
10. zA and SE are solved in single loops.
11. The zA and SE grid are written to the "Design" worksheet.
12. The matrices may be written to the "SV" worksheet for numerical checking.

When alternative data is used, some changes must be made manually. However, the code may be easily extended with the use of data entry by forms, so that manual entry for alternative data may be made at the worksheet level.

Numerical check

The trend analysis can be checked by worksheet calculation, and numerically checking some results. This involves the following steps for, say, the linear trend:

1. [Insert, Worksheet], double click on the name tag and rename to "LTrend".
2. Copy the x, y, z data from worksheet "Data" to worksheet "LTrend", columns A:C.
3. Type x^2 in Cell D1 and Formula = = A2*A2 in Cell D2.
   Type y^2 in Cell E1 and Formula = = B2*B2 in Cell E2.
   Type xy in Cell F1 and Formula = = A2*B2 in Cell F2.
   Type xz in Cell G1 and Formula = = A2*C2 in Cell G2.
   Type yz in Cell H1 and Formula = = B2*C2 in Cell H2.
4. [Edit, Fill, Down, D2:H86].
5. Formula = = count(A:A) in Cell I1; Formula = = sum(A:A) in Cell J1;
   Formula = = sum(B:B) in Cell K1; Formula = = sum(D:D) in Cell J2;
   Formula = = sum(E:E) in Cell K3; Formula = = sum(F:F) in Cell K2;
   Formula = J1 in Cell I2; Formula = K1 in Cell I3; Formula = K2 in Cell H3.
6. Formula = = sum(C:C) in Cell M2; Formula = = sum(G:G) in Cell M3;
   Formula = = sum(H:H) in Cell M4.
7. Activate Cells(I5:K7) then enter Formula = = Minverse(I1:K3) and press [Ctrl, Shift, Enter].
8. Activate Cells(M5:M7) then enter Formula = = Mmult(I5:K7, M1:M3) and press [Ctrl, Shift, Enter].

Cells M5:M7 should contain the linear b0, b1 and b2 coefficients, and may be checked with the code by using the Debug facility in the "LinearTrend" subroutine.

3.1.4 Aspects of semivariogram theory

The detailed theory of regionalised variables is given, for example, in Kitanidis (1997) and Webster and Oliver (2001). Major aspects are now summarised.

Covariance-semivariance relation

The variance of the difference between 2 variables can be obtained in the same way as a quadratic equation (Wellmer, 1998). Thus the variance (σ^2) of the difference between a variable z_i at position i and z_{i+h} at position (i + h) in a random field is given by:

$$\sigma^2(z_i - z_{i+h}) = \sigma^2(z_i) + \sigma^2(z_{i+h}) - 2[\sigma(z_i) * \sigma(z_{i+h})]$$

Since z_i and z_{i+h} are from the same population, and the product term is the covariance of the 2 variables, then:

$$\sigma^2(z_i - z_{i+h}) = 2\sigma^2(z_i) - 2\mathrm{Cov}[z_i , z_{i+h})]$$

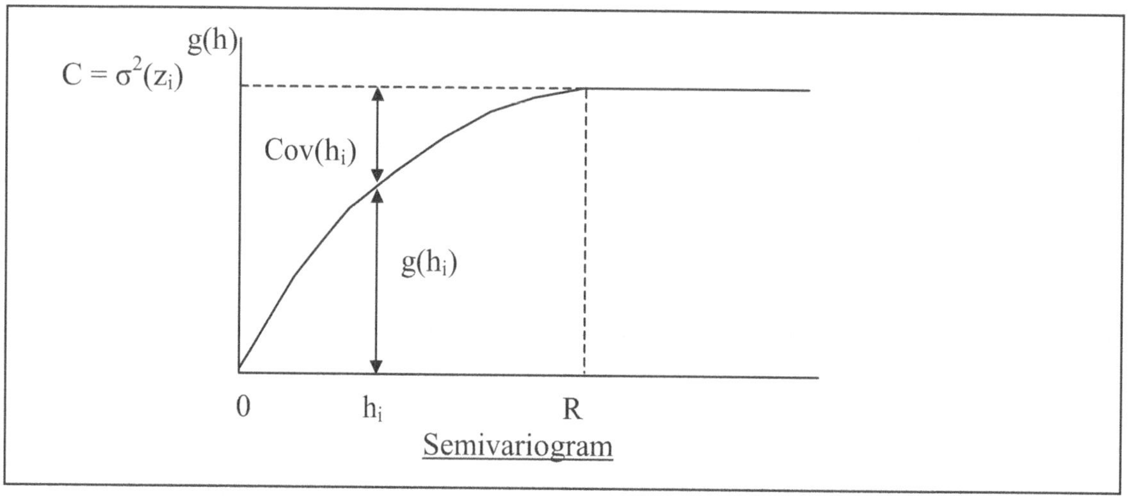

Figure 3.1.3: The relation between semivariance and covariance

Half of this difference is defined as the semivariance, thus:

$$g(h) = \sigma^2(z_i) - Cov[z_i , z_{i+h})]$$

Figure 3.1.3 shows the relation with the covariance of the regionalised variables. For h > R, the covariance is zero, and the variables are then statistically independent. The semivariance is accordingly constant, and becomes the sill value (C). For h = 0, g(h) = 0:

$$\sigma^2(z_i) = Cov[z_i , z_i)]$$

The sill value also represents the variance of the random variable. For this reason, the "Data variance" is written to the "Run" worksheet to check the SV sill values. The above considerations indicate that a covariance function may be used in place of the semivariance. It has been argued that this is more advantageous, since the covariance matrices are more easily inverted. However, a test with MS Office 2000 showed that covariance matrices showed the same inversion limitations as semivariances.

Stationarity

A continuous surface may be represented by Fourier analysis as a series of sinusoidal and cosinusoidal functions. Figure 3.1.4 depicts two simple sinusoidal surfaces, one of which is trended. A region or domain of 4 cycles is shown for this surface of z_i samples. If the first statistical moment of the data is constant, then the surface is defined as first order stationary, thus:

$$1/n \; {}_n\sum z_i = \text{constant}$$

If the surface varies in both the x and y directions, then the above condition implies that it is isotropic i.e. there is no trend in any direction.

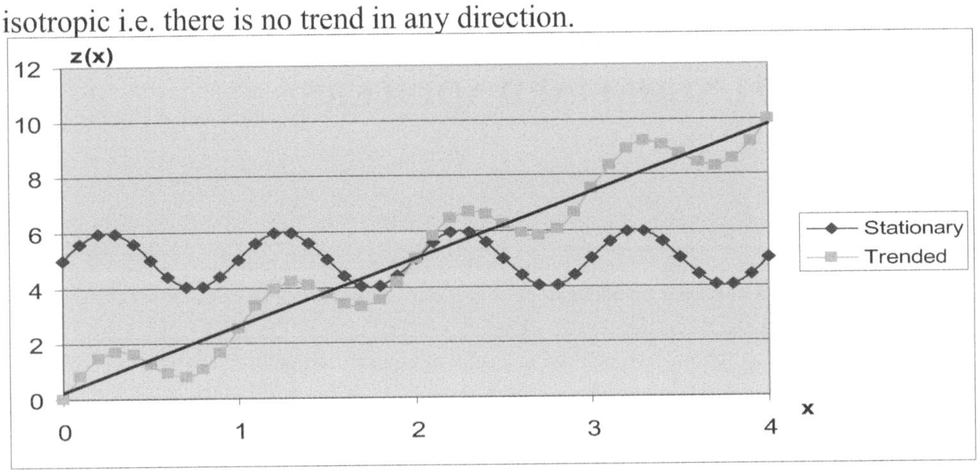

Figure 3.1.4: A stationary and trended surface

If the 2nd statistical moment of the data is a unique function of the distance (h) between sample points (x, y), then the surface is defined as 2nd order stationary, thus:

$$\mathrm{Cov}[z_i, z_{i+h})] = \mathrm{Cov}(h)$$

This means that the variables within a range (h < R) are statistically related by h.

Full stationarity is achieved if all moments of the cumulative distribution function are unique functions of h, thus the 2nd order condition is also called weak stationarity.

Even if the region is detrended as illustrated above, it is inevitable that it will become non-stationary as it is expanded. Matheron (1963) recognised this problem, and proposed the following "Intrinsic hypothesis". He assumed the sample mean was constant but unspecified, and depended only on the smaller distances (h) between sample points, thus:

$$1/n \;_n\sum z_i - 1/n \;_n\sum z_{i+h} = 0$$

It therefore takes less information to evaluate this condition, since the numerical mean is not required. Furthermore, the covariance condition was replaced by the semivariance, thus:

$$1/2n \;_n\sum (z_i - z_{i+h})^2 = g(h)$$

Again, the use of the semivariance requires less sample information, since the covariance function requires the extra estimate of the variance.

The intrinsic model is thus more general than the stationary model. However, not all intrinsic semivariance functions are stationary.

3.2 CONFINED AND UNCONFINED AQUIFERS

3.2.1 Summary

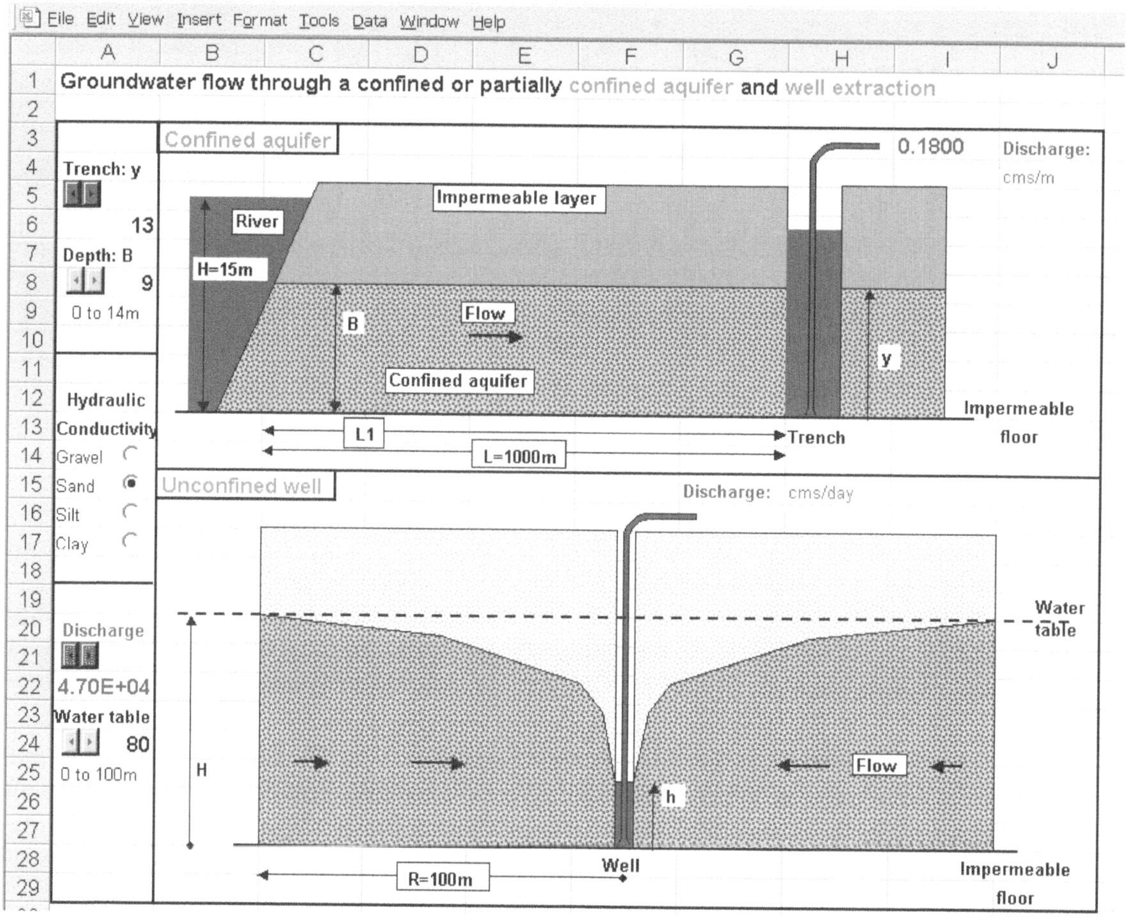

Figure 3.2.1: The “Run” worksheet for the “AQF.xls” project

These examples illustrate one dimensional, steady groundwater flow in the horizontal direction through a confined, partially confined and unconfined aquifer (Figure 3.2.1).

The confined aquifer is bounded at the top and bottom by impermeable layers. Water is provided by an adjacent river that has a constant elevation (H). A trench running parallel to the river is maintained at any desired level (y) by pumping out water at a constant discharge, Q (called “dewatering”). When y is less than the aquifer depth (B), the aquifer becomes partially confined and a different analysis for computing the pumped discharge is used.

The unconfined aquifer is bounded by a bottom layer only. Water is provided by an undefined source that maintains a constant water table (H) at a radius, R, from a pumped well. The well discharge lowers the aquifer water level (called the "drawdown"), and the drawdown profile is computed. In 3 dimensions, the aquifer water surface is called a "cone of depression".

3.2.2 Code operation

The control panel for this project (left hand side of the "Run" worksheet) is divided into 3 areas. The top area controls the adjacent confined aquifer, the middle area controls the hydraulic conductivity (K) of both aquifers, and the bottom area controls the adjacent unconfined aquifer.

For the confined aquifer, the river level (H = 15m) and trench distance from the river (L = 1000m) are fixed. This leaves the trench water level (y = 0 to 15m) and the aquifer thickness (B = 0 to 14m) as variables that may be controlled by the relevant Spin buttons. Note that y is incremented in decimals, whereas B is set to increment in integers. These may be changed at will.

For the unconfined aquifer, only the radius (R = 100m) to the undisturbed water table is fixed. This leaves the well discharge (Q) and the water table level (H = 10 to 100m) that may be controlled by the relevant Spin buttons. Note that the drawdown has a maximum at the well bottom, and there is therefore a maximum Q for each H condition. This maximum Q is set by the "Max" subroutines in the "Well" module, as described below.

Four Option buttons are used in the hydraulic conductivity area to represent 4 ranges of sediment size. The associated K values are 100 m/day, 10 m/day, 0.1 m/day and 0.001 m/day for gravel, sand, silt and clay unconsolidated sediments. One of these values is selected by a multiple If statement at the beginning of each aquifer subroutine.

3.2.3 Theory

Confined aquifer

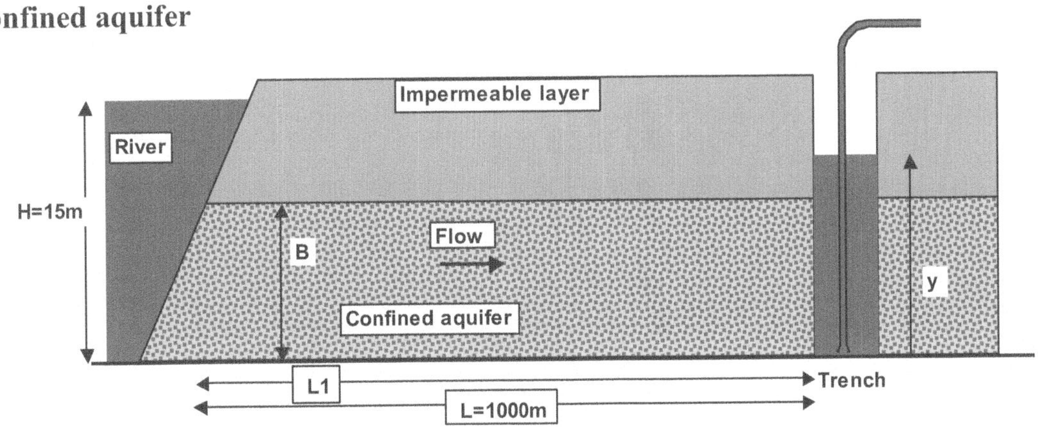

Figure 3.2.2: Confined aquifer

Groundwater flow is usually analysed using Darcy's law, which was derived in 1856. The law states that the velocity of flow (V) through a porous media is directly proportional to the hydraulic gradient (dh/dL, the gradient of the piezometric head), thus:

$$V = -K\ dh/dL \qquad \text{m/day}$$

The constant K is called the hydraulic conductivity, and is mainly a function of the sediment size and porosity when V is small. Water is pumped from the confined aquifer at constant discharge to maintain a constant level in the trench (Figure 3.2.2), thus:

$$q = -KB\ (y - H) / L \qquad m^2/\text{day}$$

where q is the discharge per unit width along the trench.

Partially confined aquifer

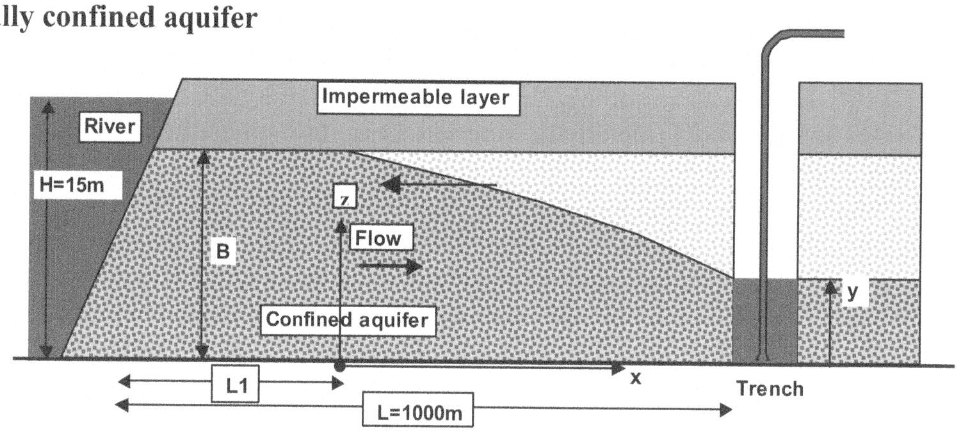

Figure 3.2.3: Partially-confined aquifer

When y < B, the groundwater surface reaches the top of the aquifer for a distance L_l from the river and then begins to drawdown (Figure 3.2.3). For this aquifer section, q is given as:

$$q = -KB\ (B - H) / L_l \qquad m^2/\text{day}$$

since the piezometric head at L_l is B. For the unconfined surface, the velocity must increase according to drawdown gradient, dz/dx, thus:

$$q = -Kz\ dz/dx$$

where the origin of coordinates is as shown. Integration of qdx from dx = L_l to L and –Kzdz for dz = B to y then gives:

$$q = K\ (B^2 - y^2) / 2(L-L_l)$$

and thus the drawdown profile is parabolic. Since q is constant, the section discharges may be equated to solve for L_l, and a final estimate for q is given (Roberson et al, 1997) as:

$$q = K/(2L)\ [2B(H - B) + (B^2 - y^2)]$$

Note that the diagram horizontal and vertical scales differ, and dz/dx is very small. Thus the one dimensional flow assumption remains valid.

Well pumping from an unconfined aquifer

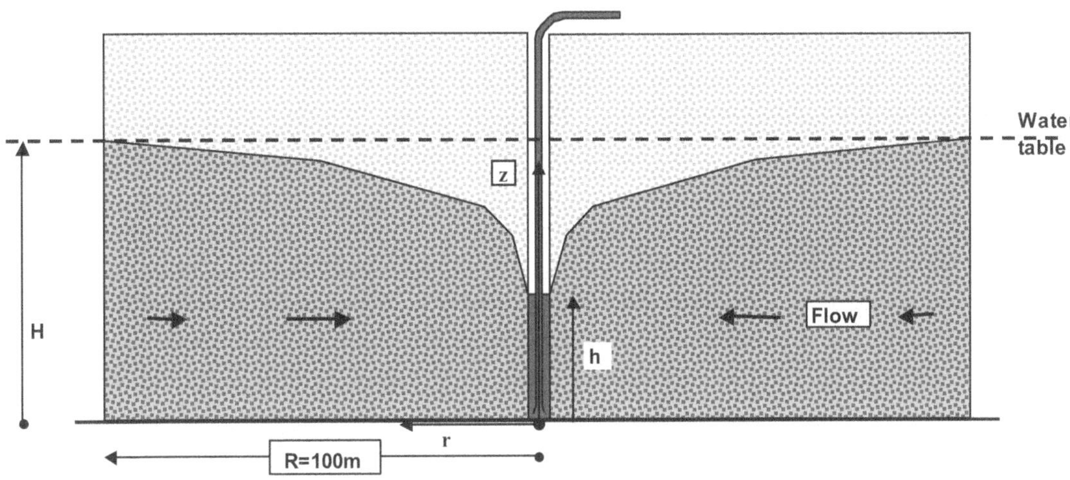

Figure 3.2.4: Unconfined aquifer

Unconfined aquifers are usually shallow and located near the ground surface. Applying Darcy's law to Figure 3.2.4, the volumetric discharge becomes:

$$Q = 2\pi Krz\ dz/dr \qquad m^3/day$$

Integrating Q dr for dr = r to R and $2\pi Krz$ dz for dz = z to H gives:

$$Q = \pi K\ (H^2 - z^2)\ /\ \ln\ (R/r)$$

This equation may be transformed to solve for z at any r, thus:

$$z = [H^2 - Q/(\pi K) * \ln\ (R/r)]^{0.5}$$

Note that a large well diameter is about 2 m. For the dimensions chosen in this project, z^2 becomes negative for both large Q and small r. Thus z cannot be solved. A well diameter of 4 m is therefore used in the code, to give a physical representation of the drawdown. (This unrealistic diameter cannot however be drawn to scale on the project's 100 m radial axis.) Furthermore, the maximum Q values are limited in the code to ensure positive values for z.

Again the flow is assumed one dimensional for the aquifer, and it follows that this is least applicable near the well for high Q. The units of the well thickness are deliberately chosen as centimetres (1 cm = 1 m/100) to emphasise this approximation towards one dimensionality.

3.2.4 Modules

"Aquifer"

This module consists of a main subroutine "ConfinedAquifer" which determines the unit discharge (flow per unit trench width). The subroutine is split into 2 parts by a logical If statement that determines whether the aquifer is totally or partially confined. In summary, the subroutine consists of 4 steps, as follows:

1. Declaration of variables, in which L, B, H , K and y are defined. Since y is plotted for single decimal places, it takes the SpinButton1 value and divides it by 10. The SpinButton1 range of values is accordingly multiplied by 10, since a Spin button increment can only be set as an integer. The 4, K values are determined by 4 logical If statements.

2. The fully confined aquifer is detected using the If y < B statement, and q is solved using the equation given in the above theory. The freeform shapes are also updated according to the new y value. These shapes are the trench water level, the trench water level arrow indicator, and the arrow indicator L1 (which is set equal to L). Note that these shapes are changed using the relevant node coordinates calculated from the worksheet position units.

3. The partially confined aquifer is detected using the Else part of the If statement, and q is solved using the equation given in the above theory. The confined length, L1, is computed, and (L – L1) is divided into 4 equal sections. The 5, x and y coordinates bounding these sections are then computed from the theory and converted to xx and yy coordinates of worksheet units.

4. Finally, the drawdown curve is plotted according to the computed coordinates. Both the right hand and left hand groundwater surfaces are plotted in addition to the 3 shapes that were replotted for the fully confined aquifer.

A second subroutine, "ChangeB" is necessary to recompute the position of the 4 aquifer shape objects that change with SpinButton2. Two of the shape objects are the impermeable top layers above the aquifer. A third shape object is the arrow indicator for B. The fourth shape object is the new drawdown surface, and this must be recomputed from the "ConfinedAquifer" subroutine. The recomputation is achieved by adding the subroutine name to the SpinButton2 control code.

"Well"

This module consists of 6 subroutines, of which 4 are repetitive. The main subroutine "WellDrawdown" consists of 4 steps as follows:

1. Declaration of variables, in which R, H , K, rw (well radius = 2 m) and water discharge (D) are defined. (Note that upper case Q cannot be used for this public subroutine, as lower

case q was used in the "Aquifer" module. Single letter, public variables of both lower and upper case cannot be used in the same VBA project.) Again, the logical If condition is used to determine K.

2. The well level is found using the above theory, and the radius and level are computed in worksheet units (rrw, hhw) for later plotting.

3. The drawdown surface is computed for 5 coordinates in a similar way to that used for the confined aquifer. Since the drawdown curve drops rapidly as it nears the well, the 4 sections were chosen as 0.0625R, 0.125R, 0.5R and 1.0R distant from the well centre. The physical coordinates for the curve are denoted by the r, h symbols, the worksheet unit coordinates by rr and hh symbols and the right hand well radii by the rrr symbols.

4. The freeform shapes associated with the drawdown are adjusted using the above computed data. These shapes are the left and right hand groundwater surfaces, the well level with its arrow indicator and the water table with its arrow indicator.

The "Textbox" subroutine is computed separately, as it moves the "Water Table" Text box upward or downward according to an increase or decrease in H. The code used to detect the direction consists of 3 VB statements, and is best explained by an example. Consider first an initial condition where H = 100 due to an increase in H from H = 90. The "Design" worksheet cells appear as below:

A1	A2	A3	A4	A5	A6	A7	A8	A9	A10	A11
0	0	0	0	0	0	0	0	0	0	100
B1										
90										

These values have been generated using the following abbreviated code:

[A11 = 100: B1 = A10 ***(= 90)***: IF A10 > 0 THEN A10 = 0]

where A11, for example, is the cell in the 11th row of column A. Note that cell B1 > 0. The H value is now decreased to H = 90, and the "Design" worksheet is as below:

A1	A2	A3	A4	A5	A6	A7	A8	A9	A10	A11
0	0	0	0	0	0	0	0	0	90	100
B1										
0										

These values have been generated by increasing H from H = 90 using the following code:

[A10 = 90: B1 = A9 ***(= 0)***: IF A9 > 0 THEN A9 = 0]

Note that B1 is now zero. In summary, if H has increased then B1 > 0 and if it has decreased then B1 = 0. The "Water Table" Text box is thus moved upward or downward according to the B1 value.

Finally, this module contains 4 similar "Max" subroutines that set the scrolling values of the Spinbutton3 control for each of the 4, K values. The Spinbutton3.Max value must not exceed a certain discharge, otherwise the well level will become negative. Since there is a very large range of numerical values (due to the varied K magnitudes), the incremental values given by Spinbutton3.SmallChange must also be set.

The above maximum values were simply determined for each K and H magnitude by trial and error. The incremental values were then chosen to give a maximum number of 100 increments for each range. Note that the Spinbutton3.SmallChange value must be an integer, and this restricts the number of increments at the lowest K value (i.e.clay sediment).

Numerical check

These project examples are readily checked numerically using the formulae given in the above theory.

3.3. GROUNDWATER FLOW BY FINITE DIFFERENCE

3.3.1 Summary

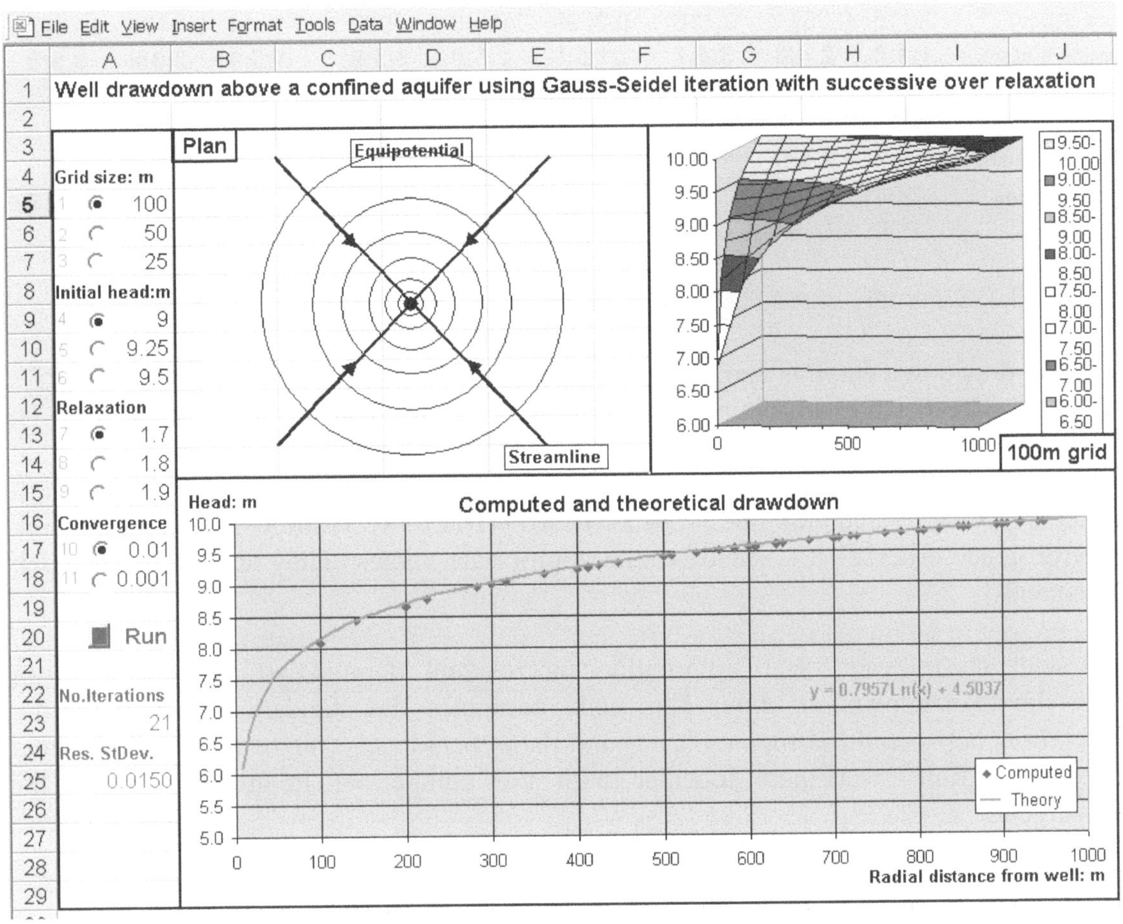

Figure 3.3.1: The "Run" worksheet for the "GWFD.xls" project

In sequence to the one dimensional groundwater well flow for an unconfined aquifer illustrated in the "AQF.xls" project, this project considers that for a confined aquifer. The drawdown curve for the former project used 5 coordinate points; it was thus discontinuous. In contrast, this project uses a finite difference method to compute the drawdown for an adjustable grid of points over the circular plan area (Figure 3.3.1). The area radius is chosen as 1000 m, and 3 square grids are used with 100m, 50 m and 25 m sides. The number of drawdown coordinates are then 82, 322 and 1273 respectively, for a quadrant.

The finite difference principle used herein involves iteratively solving the grid by computing the piezometric head at each grid node as 0.25 multipled by the sum of the 4 nearest-node heads. The iteration requires a convergence criteria. The method also includes a relaxation factor (Wang and Anderson, 1982) to speed the solution, and means to vary the initial heads chosen towards the solution. This project demonstrates how these variables influence the accuracy of the computed solution.

3.3.2 Code operation

Option buttons	1:4:7	2:4:7	3:4:7	1:5:7	2:5:7	3:5:7	1:6:7	2:6:7	3:6:7
No. Iterations	21	19	26	19	18	26	16	20	27
Res. StDev.	0.015	0.0674	0.116	0.0099	0.0188	0.1032	0.0114	0.0485	0.1347
Option buttons	**1:4:8**	**2:4:8**	**3:4:7**	**1:5:8**	**2:5:8**	**3:5:8**	**1:6:8**	**2:6:8**	**3:6:8**
No. Iterations	31	28	28	31	23	27	30	25	29
Res. StDev.	0.0085	0.0368	0.0792	0.0085	0.0127	0.0527	0.0085	0.0234	0.0878
Option buttons	**1:4:9**	**2:4:9**	**3:4:9**	**1:5:9**	**2:5:9**	**3:5:9**	**1:6:9**	**2:6:9**	**3:6:9**
No. Iterations	104	68	56	99	64	50	97	63	47
Res. StDev.	0.0096	0.0057	0.0335	0.0099	0.0057	0.0096	0.0092	0.0056	0.0253

Table 3.3.1: Summary of results for a convergence factor of 0.01

A single drawdown curve is computed and plotted when the "Run" Command button is activated. The Private subroutine for this control button selects the "Grid size" (GS) Option button which is activated using the logical If statement, and runs the relevant "Well" subroutine in the "Well" module. The "Well" subroutines detect which Option buttons are activated for the "Initial head" (IH), "Relaxation factor" (RF) and "Convergence criteria" (CC), and set the appropriate values. They again use logical If statements.

Following the relevant "Well" subroutine, the relevant "Drawdown" subroutine is run from the "Drawdown" module. This code computes the standard deviation of the differences between the computed heads and theoretical heads (the residuals), and enters it on the "Run" worksheet together with the number of iterations required for convergence.

An example for the total of 27 runs using a convergence factor of 0.01 is given in Table 3.3.1. Note that the smallest 25 m grid computations can exceed a runtime of 1 minute (using a 450 MHz CPU). The following results are noted:

1. The most accurate results (lowest line in Table 3.3.1) use the highest RF of 1.9 (Option button 9), and they require the most iterations.
2. The most accurate results approximate CC = 0.01, except for the 25 m GS (Option button 3).
3. The smallest 25 m GS has, unexpectedly, the least accurate results. However, this situation changes for a lower CC.
4. The accuracy decreases then increases with IH (Option buttons 4,5 and 6).

In summary, the computed residual standard deviation for CC = 0.01 varies from about 0.01 to 0.1 m, depending on the variables chosen for computation. This variation is considerable, and emphasises the need of these varied parameters in finite difference methods.

3.3.3 Theory

Well drawdown for a confined aquifer

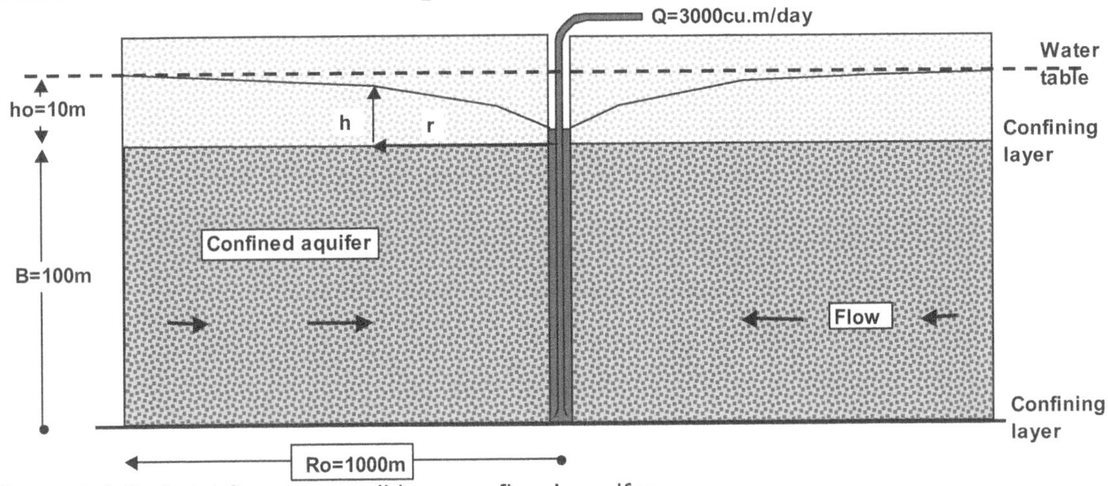

Figure 3.3.2: Axial flow to a well in a confined aquifer

Consider an aquifer which is confined above and below with impermeable layers, as illustrated in Figure 3.3.2. (The figure includes the main dimensions used for this project). Water flows horizontally and radially to the well from a limiting radius at $r = R_o$. A plan view of the flow streamlines (lines which are tangential to the flow velocity) is shown in the first drawing of the "Run" worksheet. The flow velocity is caused by the piezometric gradient normal to the streamlines, therefore the lines of equal piezometric head (equipotentials) are circles as shown in the "Run" worksheet.

The drawdown of the piezometric head has coordinates r and h, and a quadrant of the cone of depression for a 100 m grid is shown in the "Run" worksheet. Darcy's law may be applied for the pumped discharge (Q) of this steady, one dimensional flow condition:

$$Q = 2\pi KBr\, dh/dr \qquad m^3/day$$

By integrating $Q/r*dr$ from r to R_o, and $2\pi KB*dh$ from h to h_o, then:

$$Q = 2\pi KB\,(h_o - h) / \ln(R_o/r)$$

Thus:

$$h = Q/(2\pi KB)*\ln(r/R_o) + h_o$$

Using K = 6 m/day for this project with the dimensions of Figure 3.3.2, h can be solved as:

$$h = 0.7957\ln(r) - 0.7957\ln(1000) + 10$$
$$= 0.7957\ln(r) + 4.5037$$

Thus the theoretical solution for drawdown, as computed and shown on the "Run" worksheet in the lower chart, is numerically verified.

Laplace and Poisson equations

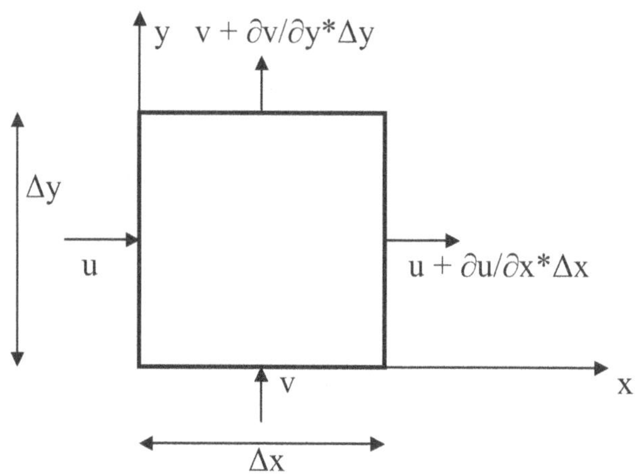

Figure 3.3.3: Control volume in a horizontal plane for vertically uniform flow

The computed solution, using a finite difference model, follows a different analytical procedure. It begins by deriving the Laplace equation for a control volume. This is extended to the Poisson equation when a point sink is added as a pumped well.

Figure 3.3.3 illustrates a small flow volume in a steady, two-dimensional flow. For the conservation of mass, the flow rate into the control volume must equal that out of the volume, thus:

$$u\Delta y + v\Delta x - (u + \partial u/\partial x * \Delta x)\Delta y - (v + \partial v/\partial y * \Delta y)\Delta x = 0$$

Hence:

$$\partial u/\partial x * \Delta x\Delta y + \partial v/\partial y * \Delta x\Delta y = 0$$

Using Darcy's equation, the velocities become $u = -K\partial h/\partial x$ and $v = -K\partial h/\partial y$. If K is equal in both directions (that is, K is isotropic), then conservation of mass becomes:

$$\partial^2 h/\partial x^2 + \partial^2 h/\partial y^2 = 0$$

This classic partial differential equation is known as the Laplace equation. If a discharge per unit area (W = Q/area) is pumped from the control volume, then the Laplace equation becomes the Poisson equation. Thus:

$$\partial^2 h/\partial x^2 + \partial^2 h/\partial y^2 = W/(KB)$$

where B is the aquifer depth as shown in Figure 3.3.2.

The finite difference method applies the Laplace equation to all grid nodes within the R_o radius, except the node at which the well is located. At the latter, Poisson's equation is used. Note that the well is not physically represented by the square and varied grids used in this project, and thus its head is omitted from the solution.

Solution by finite difference

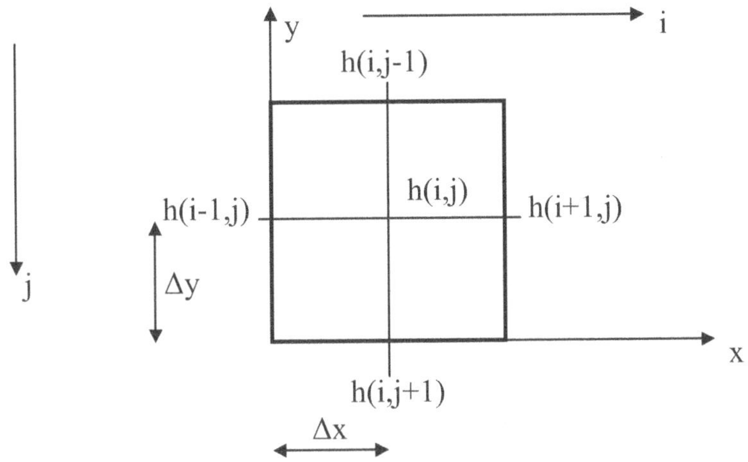

Figure 3.3.4: The 5 point operator representing finite difference approximation

Consider part of the square grid representing the well field (Figure 3.3.4). The symbol i represents a column number in the grid, and j represents a row number. The operator $\partial^2 h/\partial x^2$ is approximated by differences between nodal points, thus:

$$\partial^2 h/\partial x^2 = [(h(i+1,j) - h(i,j))/\Delta x - (h(i,j) - h(i-1,j))/\Delta x\ ] / \Delta x$$
$$= [h(i-1,j) - 2\ h(i,j) + h(i+1,j)]/\ \Delta x^2$$

Similarly:

$$\partial^2 h/\partial y^2 = [h(i,j-1) - 2\ h(i,j) + h(i,j+1)]/\ \Delta y^2$$

For a square grid, $\Delta x = \Delta y$ and thus the Laplace equation becomes:

$$h(i-1,j) + h(i+1,j) + h(i,j-1) + h(i,j+1) - 4\ h(i,j) = 0$$

or:

$$h(i,j) = 0.25\ [h(i-1,j) + h(i+1,j) + h(i,j-1) + h(i,j+1)]$$

Thus the head at a node is approximately equal to 0.25 multiplied by the sum of the heads at the 4 nearest nodes. This has become known as the “5 point operator” for finite difference approximation.

Boundary conditions

This confined aquifer problem may be solved by generating a square mesh plan above the well, and setting all the grid nodes beyond the radius, R_o, to 10 m. The latter procedure is known as setting the boundary conditions. Arbitrarily selected initial heads (IH) are then set for the nodes within R_o, and the 5 point operator is solved by starting with the known boundary heads. Such a computation is, however, excessive since the well may be considered as 4 symmetrical quadrants. Figure 3.3.5 illustrates how one quadrant only need be considered, and the boundary conditions for the x and y axes need only specify that there is no normal flow.

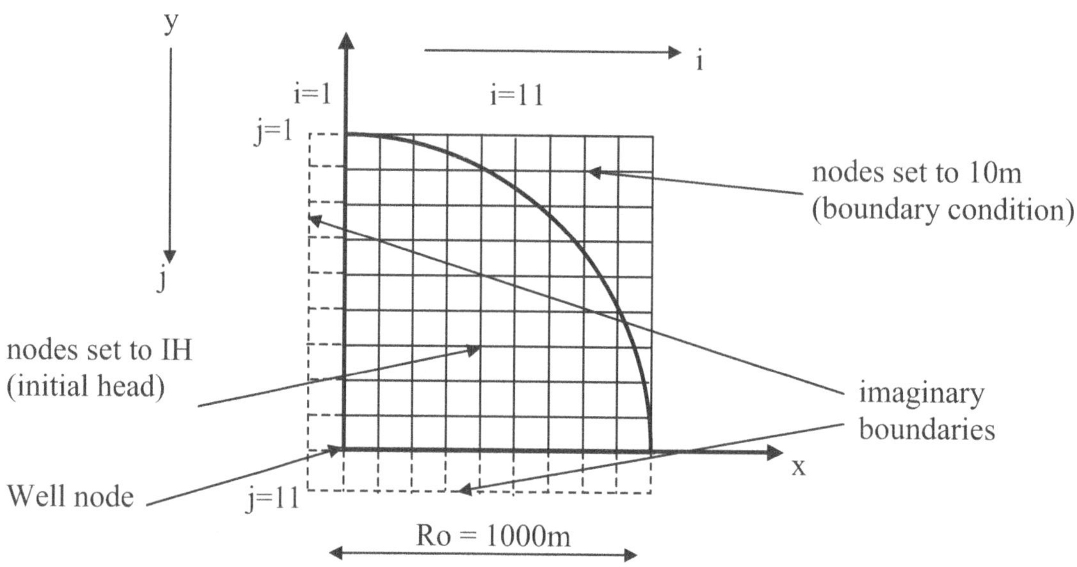

Figure 3.3.5: Finite difference model for a grid size of 100 m

Consider the y axis boundary. Since $u = -K\partial h/\partial x = 0$ for the boundary condition, we first evaluate $\partial h/\partial x$ as a finite difference approximation, thus:

$$\partial h/\partial x = h(i+1,j) - h(i-1,j)/2\Delta x$$

If h(i,j) is the head at the boundary, then h(i-1,j) is an imaginary node to the left of the y axis and h(i+1,j) is the node to the right of the axis. For the required boundary condition, these nodes must be equal. Thus the entire boundary condition is achieved by introducing an imaginary line of nodes (or imaginary boundary) whose values are set to the first column of nodes inside the boundary. The same argument applies for the x axis.

Iterative methods

Using the quadrant model with i = j = 1 at the top left node of the imaginary boundary, we could solve each non-boundary node, in any order, using the 5 point operator. The results may be written to a new matrix, and the process repeated (iterated) until a convergence criteria (CC) is met. That is, until the results for one iteration is within a certain magnitude (say CC = 0.01 m) of the next iteration. The method is named Jacobi iteration (Wang and Anderson, 1982).

A more efficient method is to solve the grid in an orderly way by moving across each row or column sequentially, and immediately replacing the old nodal value with its newly computed value. The change between 2 successive iterations (m, m+1) is called the residual, c:

$$c = h(i,j)_{m+1} - h(i,j)_m$$

By replacing $h(i,j)_m$ with $h(i,j)_{m+1}$ at every internal node, the method is "relaxing" the residuals. The relaxation amount may be increased by a factor (RF), where:

$$h(i,j)_{m+1} = h(i,j)_m + RF * c$$

It is then anticipated that the solution will reach convergence with less iterations. The optimum value for RF is usually found by trial and error using $1 < RF < 2$. This type of iteration with successive over relaxation is known as the Gauss-Seidel method (Bear and Verruijt, 1990).

3.3.4 Modules

"Well"

This module consists of 3 similar subroutines, namely "Well100", "Well50" and Well25" for each grid size. The changes made between each module are the grid size (dx) and cell references (i, j values). Note that although the advantage of object oriented programming (OOP) is to provide one object (or subroutine say) in a Class module for which the variables may be changed at will, the subroutines are repeated herein for simplicity.

Each subroutine consists of 8 steps, as follows:

1. The relevant data is entered, namely dx, Q, K, B and Ro. The product KB is called the transmissivity, T, and is also defined. The computation is to be made in the "Well" worksheet, and this is cleared for data entry.

2. The initial head (IH) is set for all the nodes in the grid. Note that an If statement is used to read the required IH value in the "Run" worksheet. Furthermore, the boundary condition and imaginary boundaries are set later.

3. The nodes nearest to r = 1000 m are determined. This subroutine iterates by row first (j, the inner loop) and then by each column (i, the outer loop). The boundary nodes are thus identified by the row number (JB) of the column in which the node distance ($d = (x^2 + y^2)^{0.5}$) is nearest to Ro.

4. The nodes nearest and greater than r = 1000 m are set to the boundary condition of h = 10 m. This is achieved by looping j from 1 to JB for each column.

5. The required relaxation factor is set from the "Run" worksheet. The finite difference method is also started by initiating an iteration counter (n), and establishing Line1 as a line number to be used in a GoTo statement.

6. The left and bottom imaginary boundary conditions are set by equating the relevant cells. Note that these conditions are reset for every iteration, to enable the fastest solution.

7. The finite difference 5 point operator is solved by looping through the nodes which are internal to r = Ro. Note that the Poisson equation for the well is set using a logical If condition, and that RF is applied in this part of the subroutine.

8. The convergence criteria is set, and if it is not met the code returns to Line1 for another iteration.

"Drawdown"

This module contains "Drawdown100, "Drawdown50" and "Drawdown25" subroutines for each grid. The modules compute the residual standard deviation in 3 steps, as follows:

1. The relevant data is entered, namely dx, Q, K, B and Ro. The computation is to be made in the "Drawdown" worksheet, and this is cleared for data entry.

2. The radial distance (d) for all internal nodes within the boundary nodes (denoted by the row indicator, JB, from the "Well" subroutines) are calculated. Note that the well node (d = 0) is eliminated for the physical reasons explained above.

3. Four columns are written to the "Drawdown" worksheet. These are the node radial distance (d), the computed head value, the theoretical head value and the residual between the latter 2 values. Finally, the residual standard deviation is computed using the Application.WorksheetFunction.StDev command, and written to the "Run" worksheet.

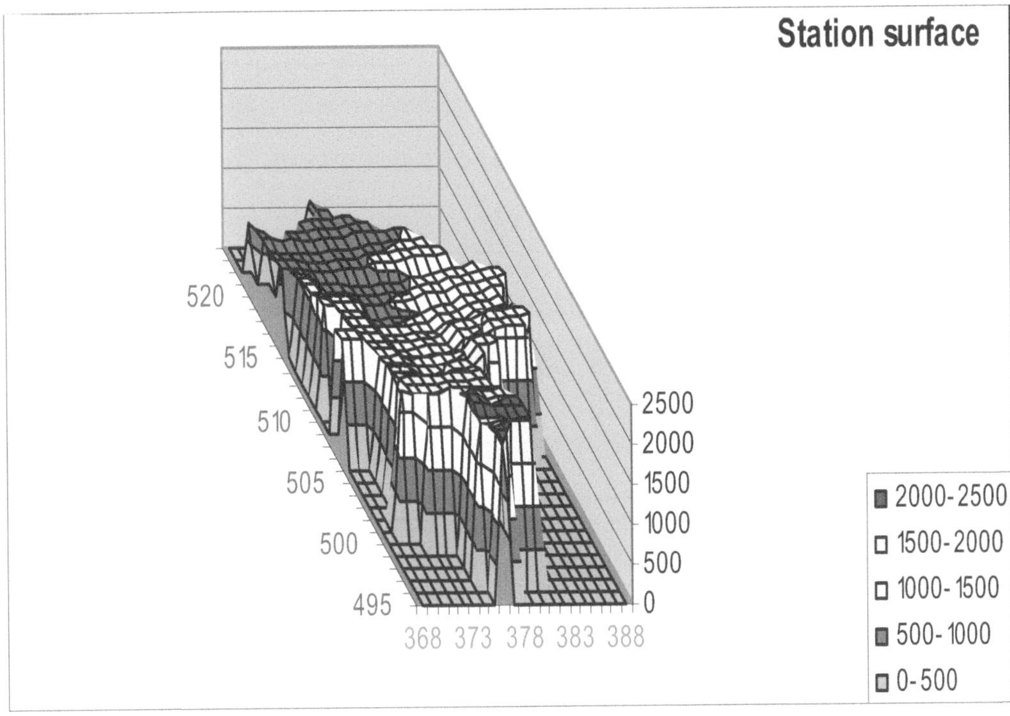

Precipitation

4.1 DESIGN LOCATION OF RAINFALL GAUGES

4.1.1 Summary

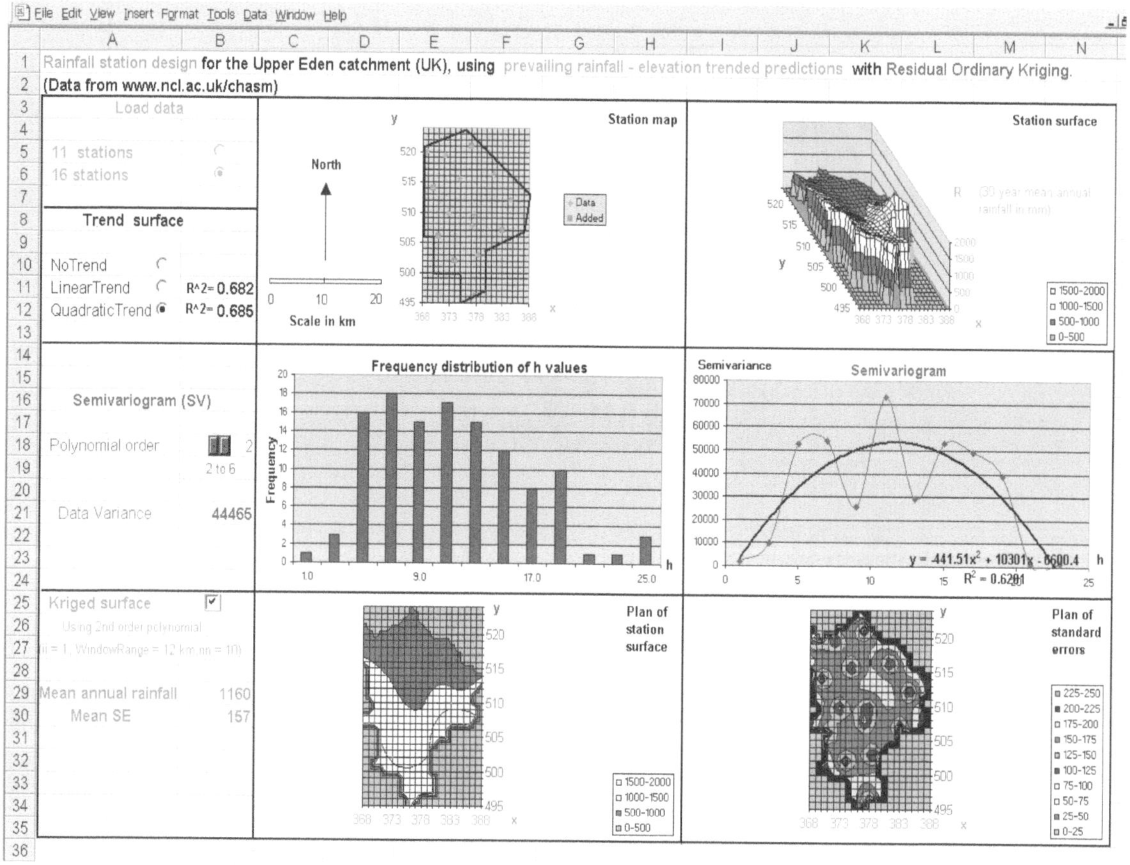

Figure 4.1.1: The “Run” worksheet for the “UEden16.xls” project

Eleven point-rainfall stations have been installed in the Upper Eden catchment (UK) for a past period of at least 30 years. The purpose of this code is to design the best locations for 5 additional stations. Residual Ordinary Kriging is first used to estimate the standard error distribution of mean annual, point rainfall (MAR) for the catchment. The design locations for 5 additional stations are then estimated from the plan of standard errors. There remains the problem of estimating MAR for the additional locations, before repeating the Kriging.

It was found for the 30 year mean annual data of the 11 stations that the highest rainfall was from the West, as expected by the Westerly prevailing winds for the area. A surface rainfall, prevailing direction-elevation trend was thus derived to predict MAR for the 5 additional stations. Residual Ordinary Kriging for the 16 stations (Figure 4.1.1) then gave a new distribution of standard errors that could be minimised by slight adjustment of the locations.

4.1.2 Code operation

Summary of results

There are 4 separated and coloured areas on the left hand side of the "Run" worksheet which contain control choices of Option buttons, Spin buttons or Check boxes. The controls are operated in sequence from top to bottom, and any non-sequential entry triggers a Message Box warning that is generated by the Sheet1 (Run) subroutines. The results are summarised below.

	11 stations			16 stations		
	No Trend	Linear Trend	Quadratic Trend	No Trend	Linear Trend	Quadratic Trend
Goodness of Fit (R^2)		0.874	0.873		0.682	0.685
Data Variance (VAR)	92843	11756	14942	136850	43570	44465
Mean annual Rainfall (MAR)	1061	1100	1098	1144	1159	1160
Mean standard error (MSE)	92	89	99	92	154	157

Table 4.1.1: Summary of "UEden16.xls" results

If the data were considered as totally random and normally distributed, the MAR with standard error is 1061 ± 92 mm for 11 stations. However, there exists a significant and similar linear or quadratic trend. Point Kriging maps this data onto grid points spaced 1 km apart, and the resulting MAR and MSE is averaged (for both trends) at 1099 ± 94 mm. The MAR increases from the random estimate since fewer stations are located in locations of higher rainfall. A similar result is noted for the 16 stations, and the addition of rainfall data at higher elevations give increased estimates for MAR of 1160 ± 156 mm.

It will be shown in the "UEden349.xls" project (below) that the Point Kriged MSE has little physical significance towards estimating a catchment (or regional) standard error. It can however be used to assess relative changes for the design of station locations.

Kriging 11 stations

The project opens with data for the 11 stations loaded into the "Data" Sheet (using subroutine "LoadData11"), and plotted according to location in the "Run" worksheet. This data is first analysed for trend. The Option Button, "NoTrend" invokes the subroutine called "NoTrend"which maps MAR for each kilometer grid point to MAR of the nearest data point. The "NoTrend" semivariogram (SV) is also computed, and it is seen that the data is non-stationary. Trending is therefore necessary.

Option Button "LinearTrend" invokes the subroutine "LinearTrend", which calculates the "goodness of fit coefficient" (R^2 = 0.874) and the SV. The latter indicates approximate stationarity for the small number of data, and a second order polynomial with sill is chosen from the Spin Button "Polynomial Order" to best model the data variance. Similarly, Option Button "QuadraticTrend" invokes the "QuadraticTrend" subroutine, which calculates the "goodness of fit" coefficient (R^2 = 0.873) and the SV. For the same reason as linear trend, a second order polynomial is chosen for Kriging. Note that the R^2 values are both high and similar; they do not indicate a trend preference.

Code operation then moves to residual Kriging. Clicking the Check box "Kriged Surface" operates the "KrigeA" subroutine, which selects the appropriate SV and plots residual levels and standard errors to the "Design" worksheet. The "KrigedSurface" subroutine is also run, and the "Station surface" and "Plan of standard errors" charts are updated in the "Run" worksheet. The code also inserts blank data into none-catchment grid points on the "Design" worksheet, so that the catchment boundary is delineated in the charts. The Kriged computation is finally summarised as "Mean annual rainfall" (MAR) and "Mean standard error" (MSE).

Rainfall direction-elevation trend

The addition of new stations for better estimates of MAR is now computed. Since the MSE for a linear trend (89 mm/km^2) is less than that for the quadratic trend (99 mm/km^2), the linear trend is used for designing new locations. The "Plan of standard errors" chart is first enlarged to a new worksheet "SE", for drawing purposes. Note that 5 major areas are present at the catchment boundary with standard errors of 100-125 mm. Five design circles which are slightly larger than the 50-75 mm Kriged data circles (to allow for future Kriging) are thus chosen subjectively. (These locations may be adjusted after Kriging to minimise MSE.) It is now necessary to assign rainfall values to the added locations.

The first estimates for rainfall values may be those of the nearest data point. However, it was observed that the 11 data sets could be reasonably separated to 6 SSE stations (South south east is 157.5 degrees of arc from North) and 5 ESE stations (East south east is 112.5 degrees of arc from North). These directions and associated stations are shown on a new worksheet, "Grid". A rainfall depth-catchment elevation correlation (R-z) was then found to exist for the separated data. Worksheet "Grid" shows this correlation with "goodness of fit" estimates of $R^2 = 0.91$ and 0.43.

Since trigonometry is based on direction angles (m) measured anticlockwise from the East direction, the SSE and ESE data were recalculated to this reference as 295.4 and 337.2 degrees of arc respectively. The direction angles were measured as the arctangent of the direction lines with the origin at the line intersection. Two options were available for extrapolating the R-z correlations to the catchment boundaries. The first option was simply to fit a quadratic trend to an R = f (m, z) surface, and extrapolate the surface. A physical disadvantage to this approach was that R in the high Northeast area would be underestimated. The alternative option of assuming R to be that of its nearest neighbour outside the data area was therefore selected. This approach is shown in 3D on a new worksheet, "Rmz".

A quadratic, direction trend surface was fitted to the physically hypothesised Rmz data using the subroutine "QuadTrendR" in the "Rainfall" module. The "goodness of fit" coefficient was $R^2 = 0.93$, and showed that the extrapolated surface was as close as the other correlations. An R value for an added station was then estimated by using the associated m and z values as shown in the "Grid" worksheet. (The z values were abstracted from a 1 km^2 Digital Elevation Model, DEM, of the catchment shown in the worksheet "DEM").

It is now assumed that the rainfall predictions are of minimal error, and the standard errors of interpolation are estimated geostatistically by examining the 16 station semivariograms for stationarity, and Kriging.

Kriging 16 stations

The new data is first loaded using the "16 stations" Option button, and the Kriging procedure follows that for the 11 stations. The "NoTrend" selection shows the SV to be non-stationary, whereas the linear and quadratic trends indicate approximate stationarity. "Goodness of fit" for the increased data is less than the 11 station data, being $R^2 = 0.682$ and 0.685 respectively. A second order polynomial is again used for Kriging. Since the polynomial intercept at $h = 0$ is small and negative, it is arbitrarily set to zero in the code (k0 = 0 in Subroutine "KrigeA").

The trended Kriged mean statistics are similar and therefore averaged herein. The MAR value increases for 11 to 16 stations, from 1096 mm/km^2 to 1156 mm/km^2; it becomes a better physical estimate since it includes the orography. Note that the 11 station data has a maximum z of 360 m (35% of the catchment is higher), whereas the added data has a maximum z of 625 m (only 2% of the catchment is higher). The MSE values also increase for 11 to 16 stations, from 95 mm/km^2 to 155 mm/km^2; it indicates the added variability of the new data.

The 16 station quadratic "Plan of standard errors", enlarged in worksheet "SE", indicates that small adjustments for the locations of the 5 added stations could improve the MSE. Several computational adjustments are however required. Firstly, a new R value must be computed for each m and z change, and entered into the subroutine, "LoadData16". Any change in R will require recomputation of the SV. The new SV polynomial variables (k0, k1, k2 and k3 in subroutine, "QkrigeA"), SV range (R in "QkrigeA") and SV sill value (C in "QkrigeA") must all be changed. These manual changes can be automated with more code, and a looping routine may be written to computationally optimise all available grid positions for minimal MSE.

4.1.3 Modules

All control operations for this code are stored as Private subroutines in the Sheet1 (Run) object. The remaining subroutines are stored in 5 modules, namely "Data", "Functions", "Krige", "Rainfall" and "Trends". The module, "Rainfall", which is additional to the "GWLevels.xls" code, deals with the prevailing rainfall, direction-elevation trending.

The following paragraphs describe the changes to this code from the previously described "GWLevels.xls" code. In several cases, the changes are small.

Sheet1 (Run) object

The operating principles for these 7 Private subroutines are similar for the "GWLevels.xls", "UEden16.xls" and "UEden349.xls" codes, and are therefore dealt with exclusively herein. Note that all the subroutines begin with the statement Application.ScreenUpdating = False. This command prevents VBA from physically changing worksheets when different worksheets are activated in a single subroutine. The Code Operator therefore sees the "Run" worksheet only, and code operation is faster.

Option buttons 4 and 5 activate the subroutines that load the 11 station or 16 station data respectively. Since this action is the start of a new Krige, the load subroutines are preceeded by clearing all "Trend" Option buttons, Check box 1 and all data in the "Run", "Design", "Data", and "SV" Sheets.

Option Buttons 1 to 3 run the "Trend" subroutines in the "Trend" module and calculate the SV in the "Data" module. To ensure there is no corruption of the required new data, all the data are first cleared in the "Design" worksheet. And if the data has been previously Kriged, Check Box 1 is finally cleared.

Spin button 1 activates the subroutine to read the required polynomial order from the "Run" worksheet, and calculate the polynomial coefficients. These commands are preceeded by a Message box warning that ensures a trend has been run.

Finally, Check box 1 calculates and plots the Kriged surface in the "Run" worksheet. If Check box 1 is clicked to be false, the unkriged trend surface is plotted. These commands are preceeded by a Message box warning to ensure that linear or quadratic (residual) trend subroutines have been run.

"Data"

"LoadData" subroutines are written in the "Data" module. The code simply sends the x, y, and z coordinates to the appropriate cells in the "Data" worksheet, and copies the data to the "SV" worksheet for subsequent analysis.

It is the "LoadData" subroutine that must be altered if alternative data is to be used. Note that the present code can use a maximum of 362 data sets. The limitation is set by the maximum worksheet row size of 65,534 and the number of independent neighbours used in Kriging (n(n-1)/2). Note that code runs faster if variables are first declared at the start of a module. If a variable is not declared, it adopts a variant type that is stored as 16 bytes. Most of the code used herein uses either integer (2 bytes) or single (4 bytes for floating point) variables. The integer variable has a maximum value of 32,767, and since this is exceeded for 256 Kriging stations, several of the integer variables must be changed in type to long (4 bytes). This necessary change has been made in the present code, and the integer variables of i, j, and k are thus dimensioned as long.

Note also that the easiest way to prepare a data set is by using the worksheet, and copying and pasting cells to the VB module. The tedious chore of writing

[Cells(i, j).Value =]

is facilitated using the worksheet [Edit, Fill, Series], [Edit, Replace] and [Edit, Fill, Down] commands.

The subroutines “KrigedSurface” and “UnKrigedSurface” are also located in the “Data” module. The former subtracts the Kriged residual elevations from the Trended elevations, whilst the latter does the reverse. Note that the former also adds the geographic boundary to the data map, by copying the DEM blank cells (denoted as a blank string, “”, in data cells of the “DEM” worksheet). Note that MAR and MSE are finally calculated using the worksheet code:

```
Application.WorksheetFunction.Average
```

“Functions”

This module is similar to that of the “BlockKrige.xls” and “GWLevels.xls” code except that Point Kriging is simply used by setting ii = 1. In similarity with the “GWLevels.xls” code, a polynomial SV is used with variables passed from the “KrigeA” subroutine.

“Krige”

The “KrigeA” subroutine begins with a multiple If statement for selecting the appropriate polynomial SV variables. The point grid is then set as a double loop, and gA is directly set equal to zero to speed compilation. Since a maximum of 16 station points are Kriged, there is no need to restrict the SV matrix with a Window range (WR) or number of nearest neighbours (nn). WR is thus set equal to the catchment diagonal dimension (40 km) and the nn statement is deleted.

“Rainfall”

The quadratic direction-elevation trending of rainfall data is identical to the Kriging, quadratic trending code, except that different input data are used. The subroutine “QuadTrendR” is therefore a duplicate of “QuadTrend” except that data is read from and written to the “Rmz” worksheet.

“Trends”

This module is identical to that for the “GWLevels.xls” code, except for the grid change.

Numerical check

This code involves Point Kriging and two types of Trend analysis. The Point Kriging of 11 stations will require the solution of a 12 * 12 SV matrix and the latter is best checked by writing all matrices to the “Design” worksheet. The “KrigeA” subroutine includes code for this to be done. The “OKrige.xls” code then details the necessary steps for solution.

Similarly, the “Trend” subroutines may be checked by worksheet commands, as described in the “GWLevels.xls” code.

4.2 MEAN CATCHMENT RAINFALL AND ERROR

4.2.1 Summary

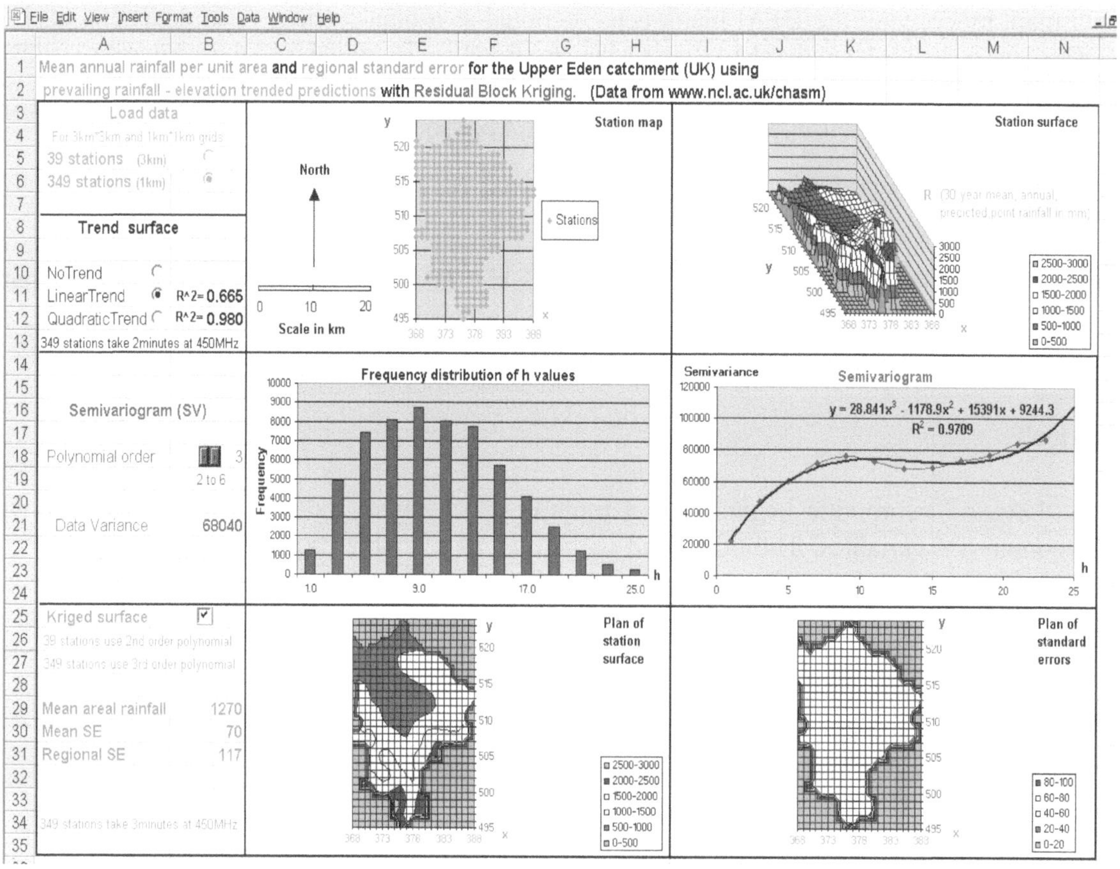

Figure 4.2.1: The “Run” worksheet for the “UEden349.xls” project

Since rainfall is generally measured at several points in a catchment, a major objective is to interpolate the measurements for estimating the mean rainfall per unit area with standard error. The result may then be used as input for modelling groundwater flow and streamflow for the catchment. This code estimates the mean, annual, areal rainfall (MAR) for the Upper Eden catchment (UK) using residual Block Kriging with a 1 km * 1 km grid (Figure 4.2.1). The rainfall for each grid point is estimated from the prevailing rainfall-elevation trend described in the “UEden16.xls” code, and a 1 km^2 Digital Elevation Model (DEM).

The Block Kriging uses 349 data points, and since this takes several minutes to run on a slow computer, the example is also accelerated and approximated to a 3 km * 3 km grid. An estimate for the MAR with the catchment or regional standard error (RSE) is computed.

4.2.2 Code operation

Summary of results

There are 4 separated and coloured areas on the left hand side of the "Run" worksheet, which contain control choices of Option buttons, Spin buttons or Check boxes. The controls are operated in sequence from top to bottom, and any non-sequential entry triggers a Message Box warning that is generated by the Sheet1 (Run) subroutines. The results are summarised below.

	39 stations			349 stations		
	No Trend	Linear Trend	Quadratic Trend	No Trend	Linear Trend	Quadratic Trend
Goodness of Fit (R^2)		0.694	0.698		0.665	0.980
Data Variance (VAR)	139512	42709	39065	191400	68040	74096
Mean areal rainfall (MAR)	1251	1267	1268	1270	1270	1269
Mean standard error (MSE)	60	87	84	23	70	72
Regional standard error (RSE)		150	158		117	123

Table 4.2.1: Summary of "UEden349.xls" results

If the data were considered as totally random and normally distributed, the MAR are 1251 mm/km^2 and 1270 mm/km^2 for 39 stations and 349 stations respectively. However, there exists a significant and similar linear or quadratic trend. The residual data variance is then much less than the variance of the original data. The MAR for 349 stations is similar for all estimates, and this provides a useful numerical check for the computations. Data variance is however higher for the 349 stations since more stations of varied rainfall are considered. The MSE for trended estimates reduce slightly with smaller grid area; they do not however represent the standard error of the mean values (as discussed later). The regional standard errors give a better estimate for the MAR, and decrease from about 154 mm/km^2 to 120 mm/km^2 with reduction in grid area.

Kriging 39 stations

The code used herein is similar to the "GWLevels.xls" code, except that a different grid is now being used and the rainfall data is first obtained by extrapolation from a prevailing rainfall-elevation trend.

The "Run" worksheet opens with data for the 39 stations loaded into the "Data" worksheet (using subroutine, "LoadData39"), and plotted according to location in the "Run" worksheet. This data is first analysed for trend. The Option button, "NoTrend" invokes the subroutine, "NoTrend", which maps MAR for each grid area to MAR of the nearest data area. The "NoTrend" semivariogram (SV) is also computed, and it is seen that the data is non-stationary. Trending is therefore necessary.

Option button, "LinearTrend" invokes the subroutine, "LinearTrend", which calculates the "goodness of fit coefficient" (R^2 = 0.694) and the SV. The latter indicates approximate stationarity for the small number of data, and a second order polynomial with sill is chosen for modelling since a third order gives a negative and unrealistic nugget effect. Similarly, Option button, "QuadraticTrend" invokes the subroutine, "QuadraticTrend", which calculates the "goodness of fit" coefficient ($R^2 = 0.698$) and the SV. For the same reason as linear trend, a second order polynomial is chosen for Kriging. Note that the R^2 values are both fairly high and similar; they do not indicate a trend preference.

Code operation then moves to residual Kriging. Clicking the Check box, "Kriged Surface" operates the "KrigeA39" subroutine, which selects the appropriate SV and plots residual levels and standard errors to the worksheet "Design". The "KrigedSurface" subroutine is also run, and the "Station surface", "Plan of residual surface" and "Plan of standard errors" charts are updated in the worksheet "Run". Note that each 3 km grid area value is mapped to the 8 adjacent kilometre grid areas using the "MapZres9" and "MapSE9" subroutines.

The code also inserts blank data into none-catchment grid points on the worksheet, "Design", so that the catchment boundary is delineated in the charts. The Kriged computation is finally summarised as "Mean areal rainfall" (MAR), "Mean standard error" (MSE) and "Regional standard error" (RSE). (The latter is computed at the end of the "KrigeA39" subroutine.)

Kriging 349 stations

The new data is first loaded using the Option button, "349 stations", and the Kriging procedure follows that for the 39 stations. The "NoTrend" selection shows the SV to be non-stationary, whereas the linear and quadratic trends indicate stationarity. "Goodness of fit" for the increased data is less than the 39 station data for the linear trend, being $R^2 = 0.665$. It is however high for the quadratic trend ($R^2 = 0.980$) and experience has shown that a quadratic trend surface for this catchment may sometimes become unstable for a small data change. For this reason, a cubic trend surface with a saddle point may be more relevant. A third order polynomial now provides a close fit for Kriging using the "KrigeA349" subroutine.

As discussed in the previous Kriging codes, there are 3 main variables to set for Block Kriging. These are the number of nearest neighbours used (nn), the Window range (WR) of the moving neighbourhood, and the number of points (ii) used in a grid area. The nn value must be set to a maximum of about 60 for large data sets, due to the Excel instability for larger matrices. The WR value may accordingly be reduced to speed computation. The influence of nn on Kriging was first investigated using WR = 20 km and ii = 16 (Table 4.2.2). The Kriging showed that a value of nn = 8 was sufficient for both trends. A WR value was accordingly set at 10 km.

The data was then tested for various ii values (Table 4.2.3, using a 450 MHz CPU), and a value of ii = 16 was selected for optimum speed and accuracy.

Linear Trend	nn =1	nn = 2	nn = 4	nn =8	nn = 16
Mean areal rainfall (MAR)	1270	1270	1270	1270	1270
Mean standard error (MSE)	117	96	76	70	70
Regional standard error (RSE)	117	117	117	117	117
Quadratic trend					
Mean areal rainfall (MAR)	1269	1269	1269	1269	1269
Mean standard error (MSE)	123	100	79	72	72
Regional standard error (RSE)	123	123	123	123	123

Table 4.2.2 Influence of nearest neighbours (nn) on Kriging

Linear Trend	ii=1	ii =4	ii = 16	ii = 64	ii =256
Mean areal rainfall (MAR)	1270	1270	1270	1270	1270
Mean standard error (MSE)	0	86	70	66	65
Regional standard error (RSE)	0	126	117	115	114
Time for computation (minutes)	2.5	2.7	3	5	12

Table 4.2.3: Influence of area points (ii) on Kriging

4.2.3 Modules

The control operations for this code are similar to those of the "UEden16.xls" code with the main exception of using different subroutines for Kriging. The 5 modules, namely "Data", "Functions", "Krige", "Rainfall" and "Trends" are also similar and the following paragraphs describe necessary changes.

"Data"

Two sets of data are loaded by the "LoadData" subroutines, and it is these modules that may be changed to run alternative data. Note that the subroutines also delete previous data written to the "Design" and "SV" worksheets so that corruption by past data is eliminated. The subroutines therefore begin a new computation.

Separate "SV" subroutines are used for the 2 data sets, since a nearest neighbour is 3 km (the lag distance, h) for 39 stations and 1 km for 349 stations. As with most repeated subroutines, the code may be reduced to one "SV" subroutine if variables are passed as necessary. However, the Code operator is then provided with more data entry and controls.

The module ends with the "MapZres9" and "MapSE9" subroutines that complete the 3 km * 3 km grid in the "Design" worksheet. Each subroutine is simply a double loop to map the grid, and the inclusion of the 8 nearest neighbour Kriged values within the nested loop.

"Functions"

This module is similar for all Block Kriging codes, with only changes made for ii (set to 16) and the SV variables passed from the "KrigeA" subroutine.

"Krige"

The "KrigeA" subroutines begin with a multiple If statement for selecting the appropriate polynomial SV variables. As with the "SV" subroutines, each data set requires a different "KrigeA39" and "KrigeA349" subroutine due to the different grids, WR and nn settings.

"Rainfall"

The quadratic direction-elevation trending of rainfall data in the "QuadTrendR" subroutine has been detailed in the "UEden16.xls" code. Two additional subroutines are now necessary for predicting rainfall for the 3km and 1 km grid elevations, namely "TrendedR39" and "TrendedR349". A new worksheet "Rainfall" is used to hold the x, y (location) and z (elevation) data for each grid. The subroutines then calculate the direction angles of the locations (m) and the Rainfall (R) as a function of m and z using the quadratic trend coefficients in the "Rmz" worksheet.

"Trends"

This module is similar to the "UEden16.xls" code, except for the grid change. Two separate "NoTrend" subroutines are used, since the 1 km data does not require computation for nearest neighbours.

Alternative Data

The above code changes also indicate the amendments required for using alternative data. For example, if a 2 km * 2 km grid is investigated for the catchment, the following code changes and additions are necessary:

1. A new Option button named, say "87 stations", is required for the "Run" sheet.
2. A new Option button subroutine is required in the Sheet1 (Run) object. This will refer to a new "LoadData87" code. The "CheckBox1" subroutine must now include the new Option button and the new "KrigeA87", "MapZRes4" and "MapSE4" commands. And the "OptionButtons 1,2 and 3" subroutines must include the added Option button code. (Note that the "NoTrend39" and "SV39" subroutines are also suitable for the 87 stations.)
3. The "DEM" worksheet may be copied to a new worksheet, say "DEM (2)", and the new data prepared as for the 3 km grid example.
4. The new x, y, z data grid must be written to columns and copied to the "Rainfall" worksheet. An example code to do this is given in the "GridToColumn4" subroutine included at the end of the "Data" module.

5. The "TrendedR39" subroutine must now be amended to "TrendedR87" by changing the loop values (i = 88) and writing m and R to the appropriate columns on the "Rainfall" worksheet.
6. The new x, y, R data may now be prepared as a "LoadData87" input file by copying the relevant "Cells" commands from "LoadData349" to the "Rainfall" worksheet, and inserting the new data.
7. There remains the "KrigeA87", "MapZRes4" and "MapSE4" to prepare since these involve the grid change.

The above procedures can of course be automated with the aid of variable input from a VB Forms object.

4.2.4 Regional standard error

Theory

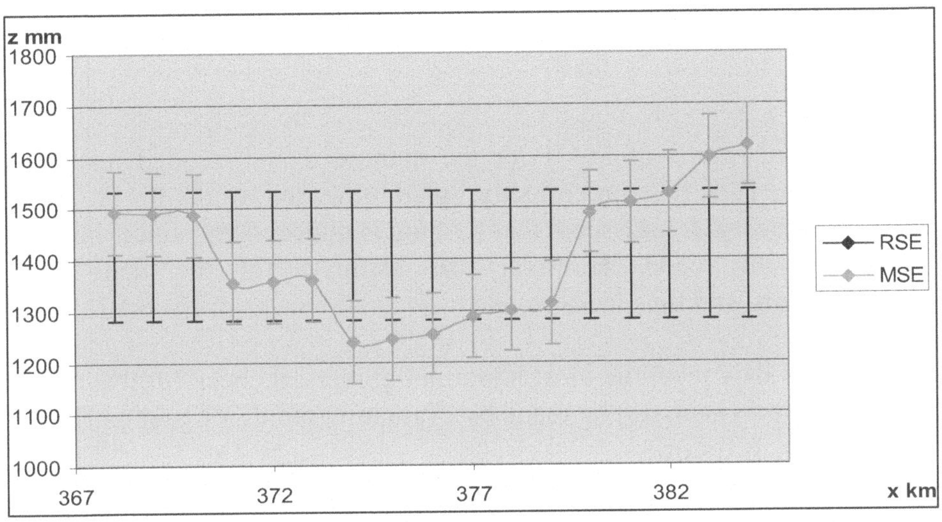

Figure 4.2.2: Regional and Mean standard error

There are two main problems for associating the mean standard error (MSE) obtained by Kriging with the standard error (SE) of the mean areal rainfall (MAR). The first is that the SE of each grid area is associated with each grid elevation, and is not representative of the ensemble (Figure 4.2.2). The second is that each grid area SE is computed from the lag distance (h) of its nearest neighbours (nn); it is therefore a dependent value. (Note that each grid area SE is equal, since it has the same configuration of nn throughout the interior of the grid. This condition does not hold at the catchment boundaries however, where the SE increases due to nn asymmetry and thus increased h values.)

It is therefore necessary to isolate each grid area (A), and estimate the SE as an indepent value (Webster and Oliver, 2001). Denoting VAR(A) as the variance of each grid area, we have:

$$VAR(A) = 2g(z_c,A) - g(A,A)$$

where $g(z_c,A)$ is the average semivariance between a point at the centre of the grid (z_c) and all other points in the area, and g(A,A) is the variance for all points in the area. Since all grid areas are equal, VAR(A) is equal throughout the catchment and the mean value is the same. The standard error is called the regional standard error (RSE) and is the square root of VAR(A). In terms of the computer code in the "KrigeA" subroutine, we therefore have:

$$VAR(A) = 2 * PSVF(z_c,A) - gAF(A,A)$$

where PSVF and gAF are the double integration and quadruple integration functions in the "Functions" module.

As illustrated in Figure 4.2.2, it is expected that the RSE will be larger than the MSE. This is confirmed by the summary of results in Table 4.2.1. Note that for nn = 1, the RSE becomes equal to the MSE (Table 4.2.2). This is because the first nearest data neighbour is at the grid point itself. Also note that for ii = 1, both MSE and RSE are zero since Point Kriging is interpolating each grid point onto itself (Table 4.2.3).

Numerical check

The almost identical results of MAR for the untrended and trended results provide a useful check that the code has run correctly. In addition, the results for nn = 1 and ii = 1 are in accord with theory. It may seem that Kriging is unnecessary, since the MAR can be estimated without the code. However, the estimate for the RSE requires the semivariogram details, and thus the SV code.

It follows from the above that the most important numerical check for this code is that of the RSE. The simplest check can be made by evaluating the PSVF and gAF functions for ii = 4 (Figure 4.2.3), as follows:

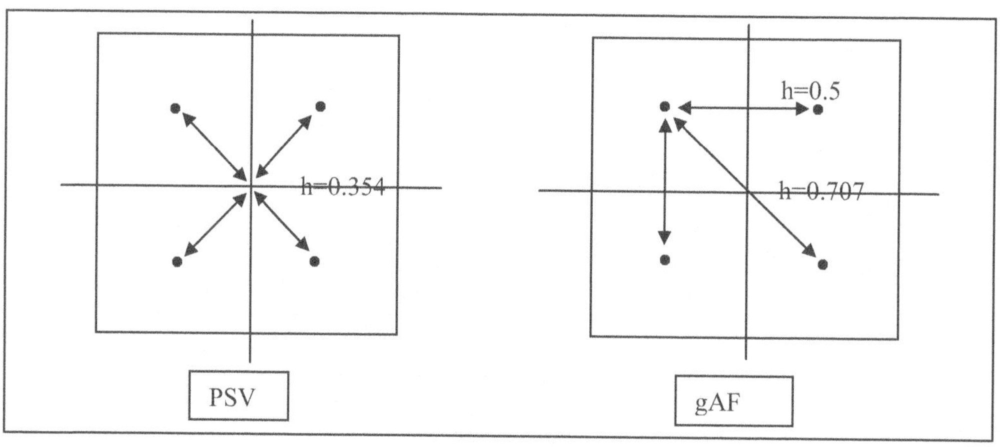

Figure 4.2.3 Calculation of the semivariogram functions for ii = 4, grid area = 1 km^2

Function PSVF is evaluated as the average semivariance between 4 subdivided area (0.5 km * 0.5 km) points from the centre of the grid area (x_c). thus:

$$PSVF(x_c,A) = 1/4\ (4 * g(\sqrt{2}/4)) = g(0.35355)$$

Using the Linear trend SV given by:

$$g(h) = 9244.3 + 15392*h - 1178.9*h^2 + 28.841*h^3$$

then PSVF evaluates to 14540. This is checked by the "TestPSVF" subroutine to be the same as the computed value for ii = 4.

Function gAF is evaluated as the average semivariance between each of the 4 subdivided area points, thus:

$$\begin{aligned} gAF(A,A) &= 1/16\ (4 * (g(0) + g(0.5) + g(0.70711) + g(0.5)) \\ &= 1/4\ (0 + 16649 + 19548 + 16649) \\ &= 13212 \end{aligned}$$

This is checked by the "TestgAF" subroutine to be the same. The RSE is therefore calculated as:

$$RSE_{ii=4} = (2 * 14540 - 13212)^{0.5} = 126 \text{ mm}$$

and is the same as the as the computed solution.

Since the RSE can be estimated without running the "KrigeA349" subroutine, the ii values may be increased beyond ii = 16 using the "TestPSVF" and "TestgAF" subroutines alone. The RSE value was then computed as 114.15 mm and 114.02 mm for ii = 256 and ii = 1600 respectively. This rapid convergence for the data indicated that the use of ii = 16 for this exercise (giving RSE = 117 mm) was adequate.

Discussion of the method

The estimation of the RSE herein assumes implicitly that the prevailing rainfall – elevation trend is without error. Although the rainfall data is highly correlated, the quadratic trend assumptions are restrictive for extrapolation to high areas outside the rainfall data boundary (particularly in the North). It would be relevant to examine data from nearby catchments, and analyse the multiple catchment rainfalls using higher order trends.

In all calculations for the semivariogram, the data has been assumed isotropic (i.e. g(h) is the same in all directions). This assumption should be tested using directional semivariograms. Note that the estimation of higher order data trends may improve isotropicity.

4.3 RAINFALL DEPTH, DURATION AND FREQUENCY

4.3.1 Summary

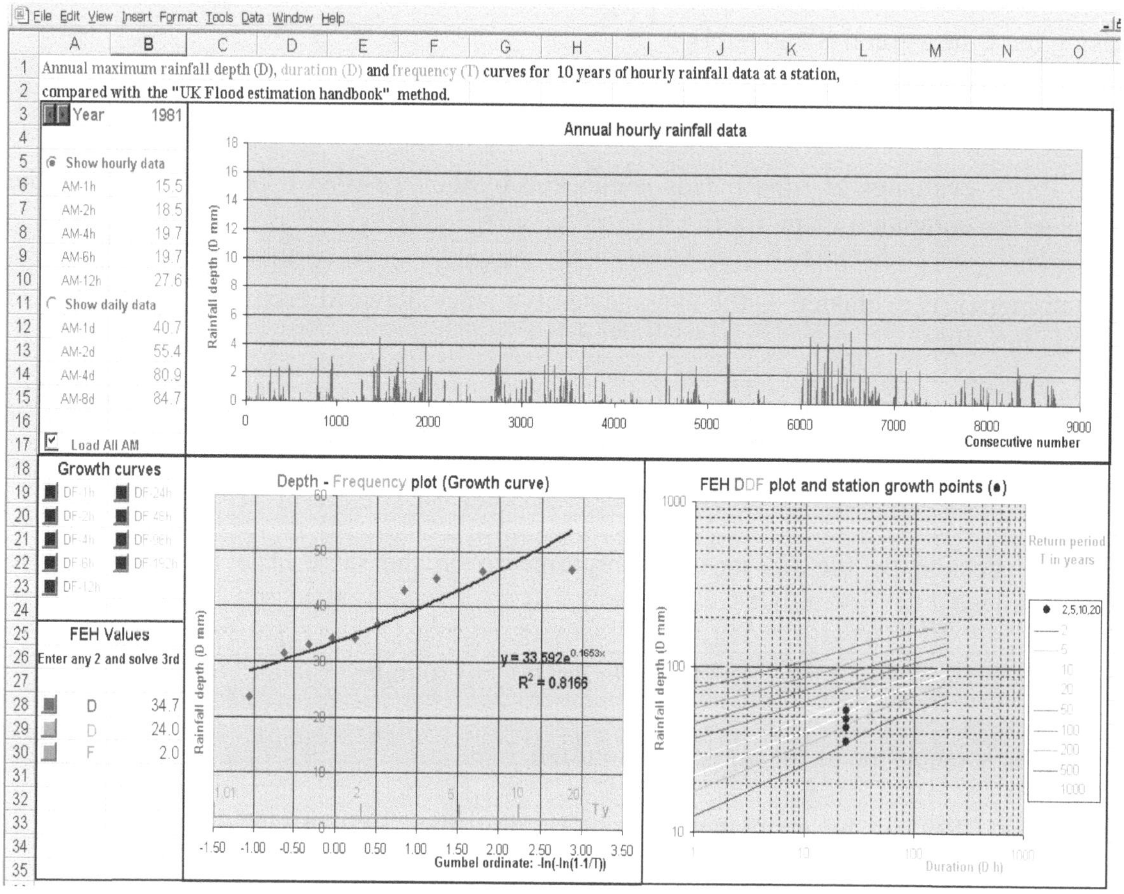

Figure 4.3.1: The "Run" worksheet for the "DDF.xls" project

This project estimates the depth, duration and frequency (measured as a return period, T) of extreme rainfall for a point location. Ten years of hourly rainfall data are examined to give annual maximum values for various durations. These values are analysed using extreme value theory to provide a depth-frequency growth curve for each duration. The extreme depths at four frequencies (corresponding to T = 2, 5, 10 and 20 years) are then plotted against duration (Figure 4.3.1).

The analysis of this single station data is compared with design curves of DDF produced by the UK Institute of Hydrology (FEH, Volume 2, 1999). The curves are derived from available rainfall data for the UK and can be used to estimate extreme rainfall depths for all points on a 1 km grid of the country. Since this project analysis is for 10 years, the comparison is made for a maximum return period of 20 years.

4.3.2 Code operation

The control panel for this project (left hand side of the "Run" worksheet) is divided into 3 areas. The top area controls the analysis of annual maximum data, the middle area controls the Growth curves, and the bottom area is a DDF solver that uses the FEH design method.

The "Year" Spin button in the top area controls several events for a particular year. It firstly selects the hourly or daily data according to the Option button activated, and enables the plot of either on the "Run" worksheet. It also controls the computation of annual maxima (AM) for the different durations, and writes these to the "Design" worksheet. Since the code runtime takes about 20 seconds for 1 year of data (using a 450 MHz CPU), a Check box is provided at the bottom of the top area that loads all the 10 years of previously computed AM into the "Design" worksheet.

The "Growth" curve Command button events will not run unless all AM data is loaded. An appropriate Message box warning is provided in the event code. The button then runs code to calculate the Gumbel reduced ordinates, and to transpose and sort the AM data so that it may be plotted on the "Growth" chart of the "Run" worksheet. The Command button finally runs the code to compute rainfall depths at the 4 selected frequencies. It assumes that the "Growth" curves are represented by an exponential distribution. Activating each Command button in sequence demonstrates how this latter assumption compares with the FEH method.

Given the grid coordinates for a station, the FEH method enables the calculation of depth, duration or frequency when any 2 of these variables are known. The estimate is for a rainfall depth at a point in space, and this depth must be adjusted if rainfall over an area is required.

4.3.3 Procedure for the single station data

Annual maxima

The following steps are used in computing the annual maxima:

1. One year of discrete 1 hour data is transposed to a single column (8760 rows for a 365 day year) for plotting on the "Run" worksheet.
2. The non-zero 1 hour data are written from the above column to a second column, and the largest value is recorded (AM-1h).
3. The values of 2 hour data (i.e. each hourly value is added to the next hourly value) are written to the second column, and the largest value is recorded (AM-2h).
4. The values of 4, 6 and 12 hour data are written sequentially to the second column, and the largest values are recorded (AM-4h, AM-6h, AM-12h).

5. Consecutive rows of 24 hour data are summed, and one year of discrete daily data is written to a single column (365 rows for a non-leap year) for plotting on the "Run" worksheet.
6. The non-zero 1 day data are written from the above column to a second column, and the largest value is recorded (AM-1d).
7. The values of consecutive 2, 4 and 8 day data are written sequentially to the second column, and the largest values are recorded (AM-2d, AM-4d, AM-8d).

Gumbel reduced ordinates

The Gumbel reduced ordinates are then calculated in the "Design" worksheet for the 10 years of annual maxima in a similar way to the "EVA.xls" project, as follows:

1. The data rank (i) from 1 to N (where N is the number of annual maxima) is defined in increasing order.
2. The Gringorten probability (P) for non-exceedence is defined as:
$$P(i) = (i - 0.44) / (N + 0.12)$$
3. The return period is calculated as $T = 1/P$.
4. The Gumbel reduced ordinate is computed as:
$$y = -\ln(-\ln(1 - 1/T) = -\ln(-\ln F)$$

The same procedure is used for the 4 frequencies representing T = 2, 5, 10 and 20 years used in the DDF plot.

"Growth" curve

The design rainfalls required for Civil Engineering projects are required to start at any time. However, the rainfalls aggregated at a station gauge start at a repetitively fixed time (hourly or daily say). It is expected that the former rainfalls (defined by FEH as a "sliding" duration event) are larger than the latter (defined by FEH as a "fixed" duration event). For example, a fixed-duration of 1 hour will not capture the highest rainfall for 1 hour during a storm. All station data must therefore be converted to sliding-duration rainfall before analyses. The following factors in Table 4.3.1 (Dwyer and Reed, 1995) were used:

Fixed-duration in hours or days	1	2	4	8	12
Sliding-duration factor for hours	1.16	1.08	1.03	1.01	1.0
Sliding-duration factor for days	1.16	1.11	1.05	1.01	1.0

Table 4.3.1: Factors for converting fixed-duration to sliding-duration data

The following sequence is used for the preparation of the "Growth" curves:

1. The annual data are first transposed, then multiplied by the factors to represent maxima which start at any time.

2. The duration columns are sorted in descending order to match the Gumbel reduced ordinates.
3. Each duration column is selectively plotted against T (and y) as a "Growth" curve on the "Run" worksheet.
4. A "least squares fit" for an exponential trendline is applied to the "Growth" curve.
5. The coefficient and power of the exponential function are used in the "Design" sheet to calculate the design depths at that duration for the 4 chosen frequencies.

The design depths for step 5 are finally plotted on the DDF chart for comparison with the FEH design method.

4.3.4 Procedure for the FEH method

Brief description

The FEH annual maximum analyses were based on UK data from 6106 daily and 375 hourly rainfall gauges; the longest records were about 100 years. The analyses involved the mapping on a 1 km grid of a "rainfall index variable" and the derivation for the same grid points of "Growth curves". Grids were prepared for rainfall durations of 1, 2, 6 and 12 hours and 1, 2, 4 and 8 days.

The rainfall index variable (RMED) was the median of annual maximum rainfalls for a particular duration, and thus corresponded to the return period (T) of 2 years (since P = 0.5). Values of RMED for the grid were interpolated between rainfall gauges using multiple regression analyses with topographic information. Rainfall growth curves at a particular duration were developed specifically for the analysis using an empirical method named "FORGEX" (FEH, Volume 2, 1999). The method involved combining data from several statistically similar sites in a region (called pooled sites) that were centred on a grid point and up to 200 km in radius. The resulting tables of RMED and growth curves were finally multiplied for each grid point to build a DDF model.

The DDF model

A logarithmic scale was used for the DDF ordinates, so that a hypothesised power law correlation for depth and duration would plot as a straight line at each frequency. Thus, if the rainfall depth (R mm) is given by $R = aD^b$, then:

$$\ln R = \ln a + b \ln D$$

For UK data, it was found that the empirical curves optimally reduced to 3 sections which were segregated by duration as $D \leq 12$ h, 12 h $\leq D \leq 48$ h, $D \geq 48$ h (Figure 4.3.2). The 12 hour separation point was physically associated with convective rainfall and the 48 hour point with frontal rainfall.

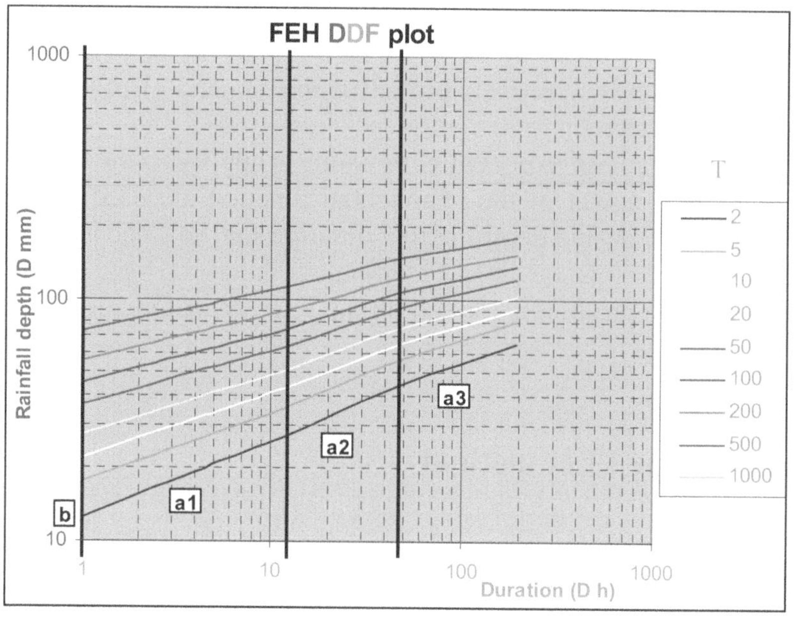

Figure 4.3.2: Slope changes for the FEH – DDF model

Each frequency curve was associated with 3 different slopes (a1, a2 and a3) and the entire line function could be represented by 6 variables (c, d_1, d_2, d_3, e and f):

for $D \leq 12$ h	$\ln R = (cy + d_1) \ln D + ey + f$
for $12\text{ h} \leq D \leq 48$ h	$\ln R = \ln R_{12} + (cy + d_2)(\ln D - \ln 12)$
for $D \geq 48$ h	$\ln R = \ln R_{48} + (cy + d_3)(\ln D - \ln 48)$

These FEH-DDF values were directly calculated as worksheet functions in the "FEH" worksheet. If a station is changed, the 6 variables are changed accordingly from the FEH tabulated data. Note that the variable values must not be deleted in the worksheet, as the links to the worksheet functions would then be broken.

Discussion

The fitting of a single straight line to a frequency curve is equivalent to fitting an exponential curve when the rainfall depth is plotted as a linear scale (as for the single station data used herein). The use of 3 straight lines therefore gives a better estimate for the UK grid data, and is one explanation for the differing design values between the single station data and the FEH method.

The main explanation for differences between the 2 methods is, of course, that the FEH uses averaged data from pooled sites rather than a single site. A problem with this method is however that the confidence limits for the predictions are uncertain.

4.3.5 Modules

The code for this project is separated into 5 modules named "Data", "Hourly", "Daily", "Growth" and "Solve". The major parts for each module are now described.

"Data"

Two subroutines are coded in this module. The first subroutine, "WriteAnnMax", copies the annual maximum data from the "Run" worksheet to the "Design" worksheet in 2 steps:

1. Year and Duration titles are written. The columns to be written in the "Design" worksheet represent the year of data and the rows represent the 9 durations.
2. Annual maximum columns from the "Run" worksheet are written to the "Design" worksheet. An If condition checks that the computed data is entered into the correct column of the "Design" worksheet.

Since the annual maximum for all the data must be computed before estimating the growth curves, the process is accelerated using the next "LoadAllAM" subroutine. The latter repeats the titling of rows and columns and then loads precomputed data as a 9 * 10 array (named AM). The data has been previously computed using the Spin button 1 control for the entire 10 years of data. Note that for longer records using alternative data, the Spin button 1 event is best written in a loop.

"Hourly"

This module begins by identifying the data required in the "DataYearHourly" subroutine, then proceeds to estimate the annual maxima for all the hourly durations. The first subroutine is in 3 steps, namely:

1. Yearly data is identified in the "Data" worksheet. The required year is read from the "Run" worksheet and converted to the relevant start row in the "Data" worksheet. Care is taken to identify the leap years.
2. Hourly data is transposed to Column "AF" in the "Data" worksheet. This enables the record to be plotted in the "Run" worksheet chart.
3. The "Run" worksheet chart scale and title are edited.

The 5 duration subroutines estimate annual maximum values in a similar way. For example, the "AnnMax2h" subroutine loops through each value in Column AF and adds the following value to its predecessor; it then writes the sum to Column AG. The worksheet function "Max" is then used to find "AM-2h" in Column AG.

The "Daily" module is similar to the "Hourly" module, except that daily values are first estimated by summing each row of 24 hours of data in the "Data" worksheet.

"Growth"

This module consists of 3 different subroutines, and 9 similar subroutines that enable the plotting of the 9 growth curves. The first subroutine, "Gumbel", computes the Gumbel reduced ordinates for the 10 years as detailed in the "Procedure for the single station data" section of this project. The second subroutine, "TransposeSlidingSort" computes the rainfall depth data for the growth curves in 4 steps:

1. Rows and columns are titled. The original AM array is transposed to an array in the "Design" worksheet, and the row and column titles are accordingly reversed.
2. Data is transposed.
3. Data is adjusted from "Fixed" to "Sliding". Data enlargement factors (Table 4.3.1) are applied for each duration.
4. The annual maximum data are sorted in descending order.

As an example for the 9 similar subroutines, the "DF1h" subroutine is described. It consists of 2 steps:

1. The Source data for the "Growth" curve is changed to the 1 hour rainfall depth column.
2. The Duration length, D (in hours), is written for the 4 selected design frequencies used in the DDF plot.

There remains the final subroutine, "MacroDDF", which reads the exponential trendline from the growth curve and calculates the rainfall depths for the 4 selected frequencies as follows:

1. The exponential formula is written to the "Design" worksheet.
2. The exponential formula is parsed. This is done by recording a macro using the Excel commands [Data, Text To columns].
3. Depths are calculated for the 4 frequencies.

"Solve"

Given any 2 of the DDF variables, the third may be solved using the FEH formula quoted above. This module therefore consists of 3 subroutines that enable the solution of each variable. Each subroutine begins by reading the 6 design variables for a curve. The values of duration are constrained between 1 hour and 192 hours, and return period is constrained between 2 years and 1000 years. The final part of each subroutine uses logical If conditions to identify the duration dependent formula.

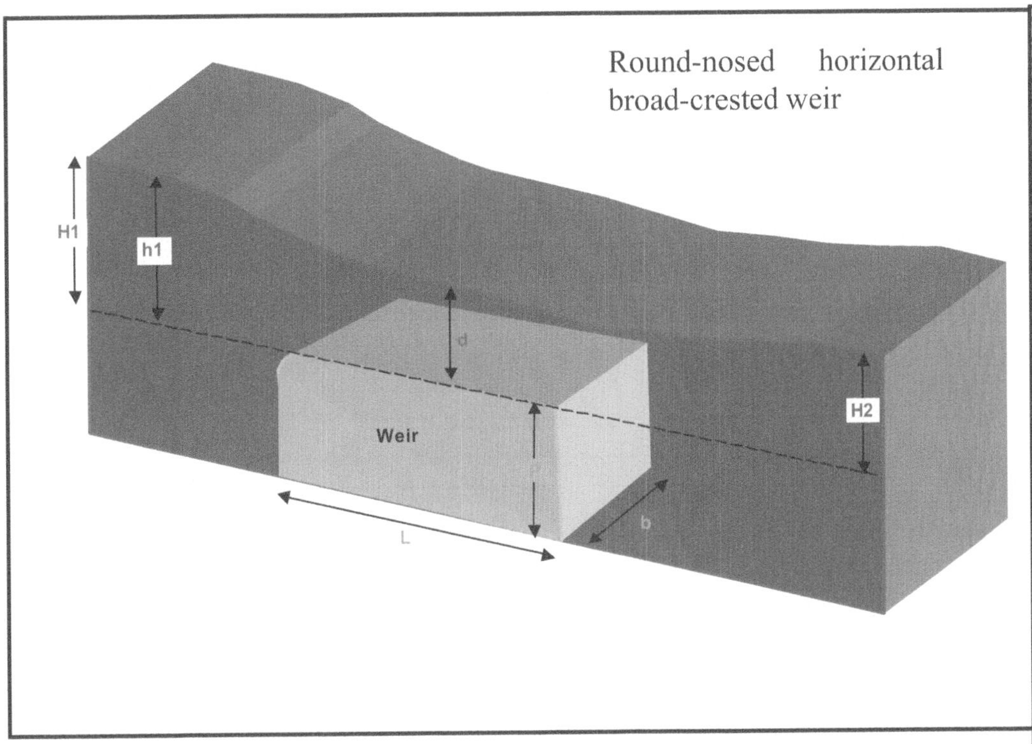

Surface runoff

5.1 ROUND-NOSED, HORIZONTAL BROAD-CRESTED WEIR

5.1.1 Summary

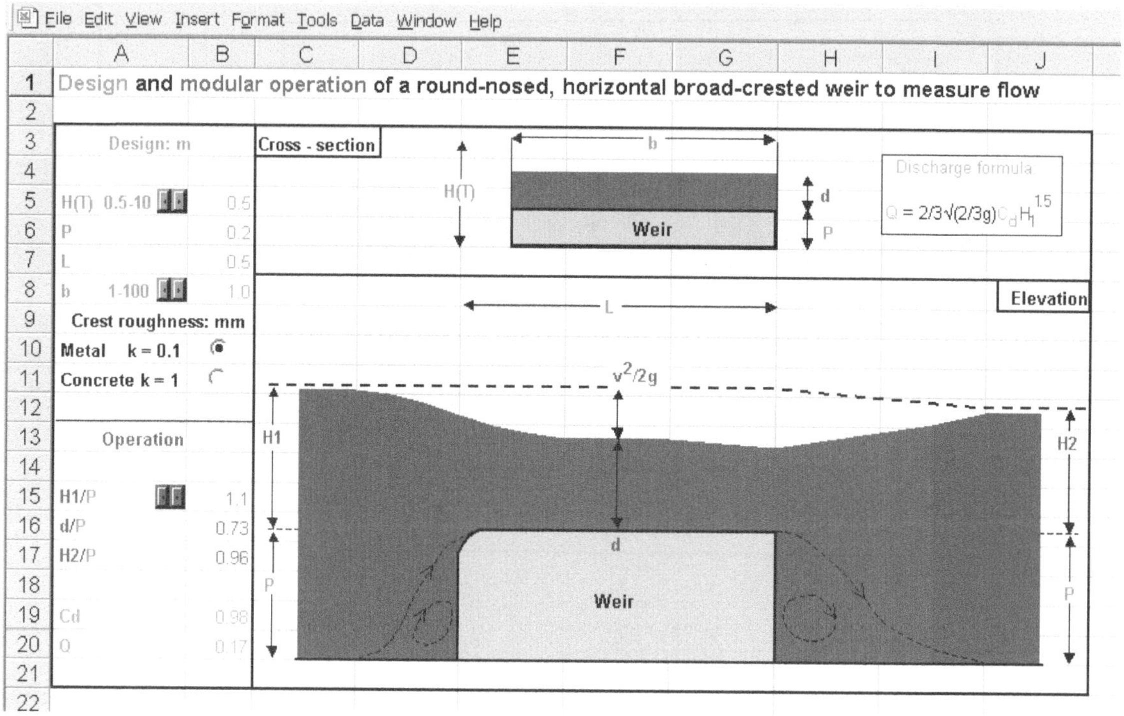

Figure 5.1.1: The "Run" worksheet for the "Weir.xls" project

This weir is used as a standard for discharge measurement, and the associated hydraulic theory behind its operation also applies to critical depth flumes (Bos, 1978). It consists of a level and horizontal crest between vertical abutments, such that its cross section is rectangular. The upstream corner is rounded in such a way that minimal flow separation occurs when the contracted flow enters the weir. The downstream corner may have a vertical drop (as in Figure 5.1.1), a downward slope, or it may be rounded.

The principle of operation is that the weir is sufficiently long as to attain critical flow conditions (of depth, d) in its section. The flow discharge (Q) may then be related to the upstream, undisturbed flow depth (H_1), and it is the latter that is monitored. The weir is designed to be independent of variations in the downstream water level (H_2). When operating in this condition, it is said to be within its "modular limits". (The procedure for designing modularity is detailed by Ackers et al (1978).)

Standards for designing this weir are given by the British Standards Institution (BS 3680, 1990). Some of these standards are demonstrated herein for this project. Note that this type of weir was originally called a "streamlined broad-crested" weir. It is now called a round–nosed, horizontal broad–crested weir to differentiate it from a rectangular profile (square-edged) broad-crested weir and other long base weirs of triangular along-stream profiles (such as the Crump weir) or triangular across-stream sections.

5.1.2 Code operation

There are two areas of operation on the control panel at the left hand side of the "Run" worksheet. The top area controls the design of the structure by means of two Spin buttons, and the lower controls the weir operation by means of one Spin button. These controls and associated output are now described.

Two field variables are available for selection by the upper Spin buttons, namely the maximum upstream head to be used in the design (H(T)), and the breadth of the stream or river (b). The variable H(T) can be estimated from an annual maximum, discharge frequency plot of a nearby station (to give Q(T) for an associated return period, T), and then finding H(T) from a rating curve (which is a plot of H against Q). For this project, the H(T) range has been arbitrarily set from 0.5 m to 10 m, so as to represent field conditions. The stream breadth, obtained by direct surveying, is also assumed to represent the weir breadth. Its range is arbitrarily taken from 1 m to 100 m.

Using BS 3680, the weir crest height (P) and length (L) may be calculated. The further variable of the physical roughness (k mm, say) of the weir crest must also be included in the design. Either metal or concrete may be used in weir construction, hence k will vary from about 0.1 mm for clean metal, to about 1 mm for concrete.

The lower Spin button operates the variable, H_1/P between the range of 0.1 and 1.5. The lower limit represents the range of experimental verification for the theory, and the upper limit is determined by the upstream Froude number (Ackers et al, 1978). Once H_1/P has been selected, d/P, H_2/P, Cd (the drag coefficient) and Q in the lower area may be computed, and the water levels simulated on the "Elevation" drawing. Note that the latter shows a fixed area of critical depth (d), for simplicity. In reality the location for d is variable; it moves upstream on the weir crest as H_1/P is decreased.

H(T)	b	k	H_1/P	Cd	Q
0.5	1	0.1	1.5	0.98	0.28
10	100	0.1	1.5	0.99	2492
0.5	1	1	1.5	0.97	0.27
10	100	1	1.5	0.99	2481
0.5	1	0.1	0.1	0.83	0.004
10	100	0.1	0.1	0.93	40
0.5	1	1	0.1	0.7	0.003
10	100	1	0.1	0.86	37

Table 5.1.1: Range of results for discharge measurement

The range of results simulated in this project are given in Table 5.1.1. Although the discharge measurement may vary from 0.003 m^3/s to 2500 m^3/s, there are alternative structures that may be more accurate and suitable for particular ranges. Furthermore, in streams with high sediment transport, deposition may occur immediately upstream and downstream of the weir. The use of a critical depth flume would then be preferable.

5.1.3 Theory

Critical depth

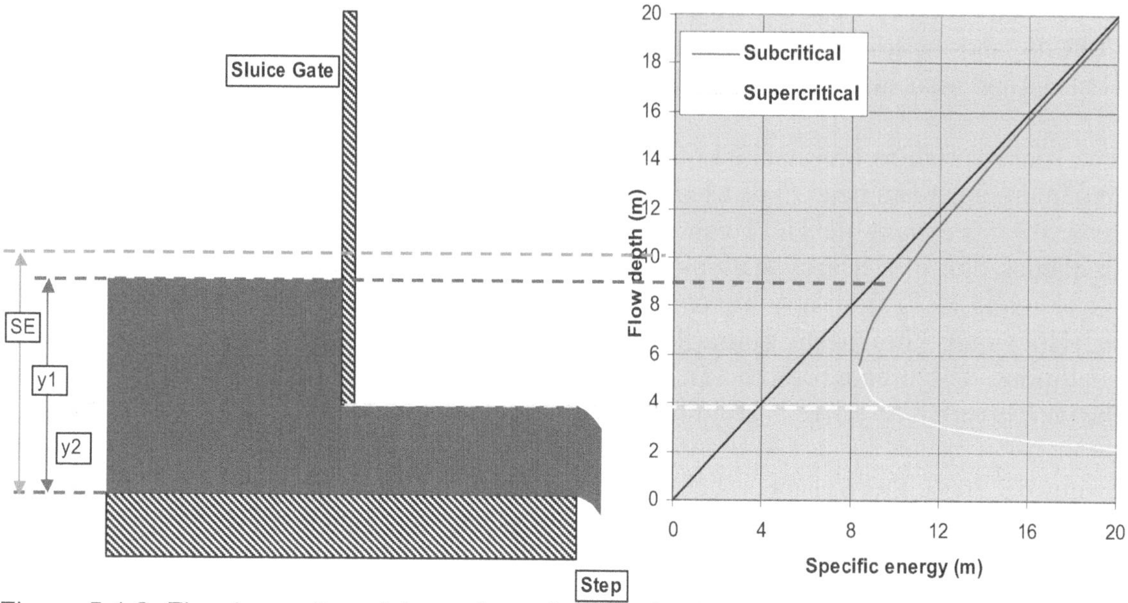

Figure 5.1.2: Flow beneath a sluice gate and over a horizontal step

Critical flow is demonstrated by considering a closed sluice gate in a horizontal channel. When the gate is opened slightly and a discharge (Q) applied, the upstream flow velocity is slow and called subcritical, whilst the downstream velocity is fast and called supercritical (Figure 5.1.2). As the gate is further opened, the flow depths become closer until they are equal at a "critical" depth.

The sum of the potential and kinetic energy head of the flow for a horizontal channel is called the specific energy (SE). It is referenced to the bed level, thus:

$$SE = y + v^2/2g$$

where y is the depth and v the velocity of the flow. By conservation of mass (continuity), Q = vA = vby (A is the channel area, b is the flow breadth), thus:

$$SE = y + Q^2/2gA^2$$

If y is plotted against SE, the curve shown in Figure 5.1.2 results for constant Q. It is seen that critical depth (y_c) occurs for minimum SE, thus:

$$d(SE)/dy = 0 = 1 - Q^2/gA^3 * dA/dy = 1 - v^2/g * b/A$$

therefore: $y_c = v^2/g$ and the Froude number, $v/(g\ y_c)^{0.5} = 1$

At y_c: $SE = y_c + v^2/2g = 3/2 y_c$

Critical flow over a weir

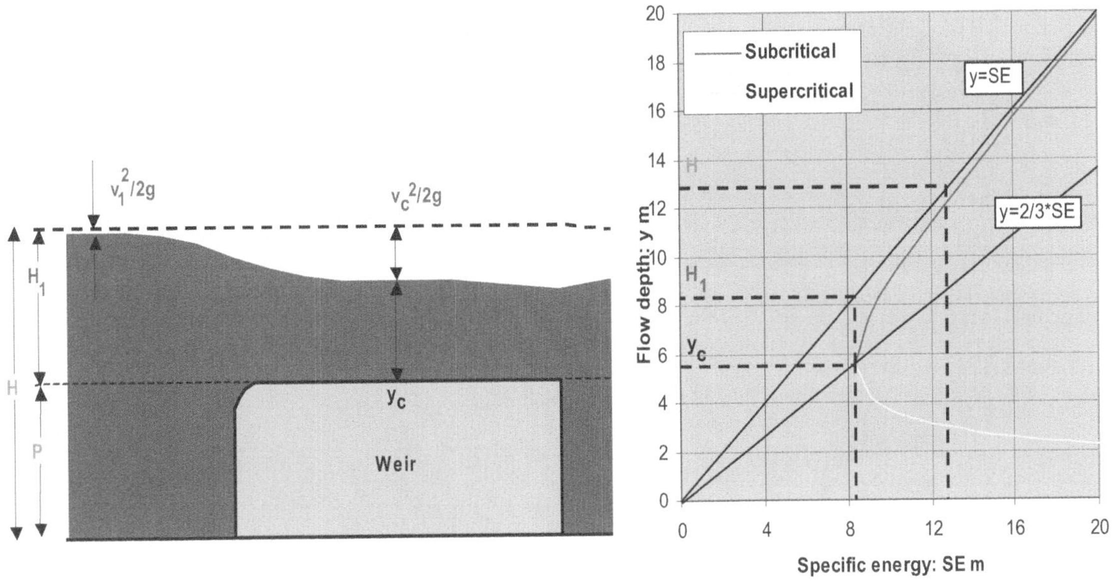

Figure 5.1.3: Flow over a horizontal–crested weir

The above considerations may be extended to the horizontal-crested weir. The upstream velocity, v_1, is usually small and may be neglected. The upstream SE is thus the energy head, H (Figure 5.1.3). As the flow accelerates at the start of the weir, the SE decreases since the bed level datum is increased. Over the weir, the SE can only decrease to a minimum value since the bed level is now constant. This value is given by the SE for critical depth (y_c) at the particular discharge.

As the flow approaches the downstream end of the weir, the verticall overfall causes a further drawdown of depth to supercritical flow (for which $y \approx 0.71y_c$). Beyond the weir, the tailwater level is increased back to subcritical flow by means of a hydraulic jump.

If the crest level (P) is used as a common datum for both upstream and weir flow, the critical depth may be approximately expressed in terms of the level, H_1, thus:

$$y_c = 2/3 * SE \approx 2/3 * H_1$$

This expression neglects friction losses within the flow. Using this common datum again, then

$$H_1 = y_c + v_c^2/2g$$

Since the discharge per unit breadth, $q = v_c d_c$, then the above equations combine to give:

$$q = 2/3(2/3*g)^{0.5} H_1^{1.5}$$

Thus the unit discharge may be estimated by the single measurement of upstream head.

Design standards

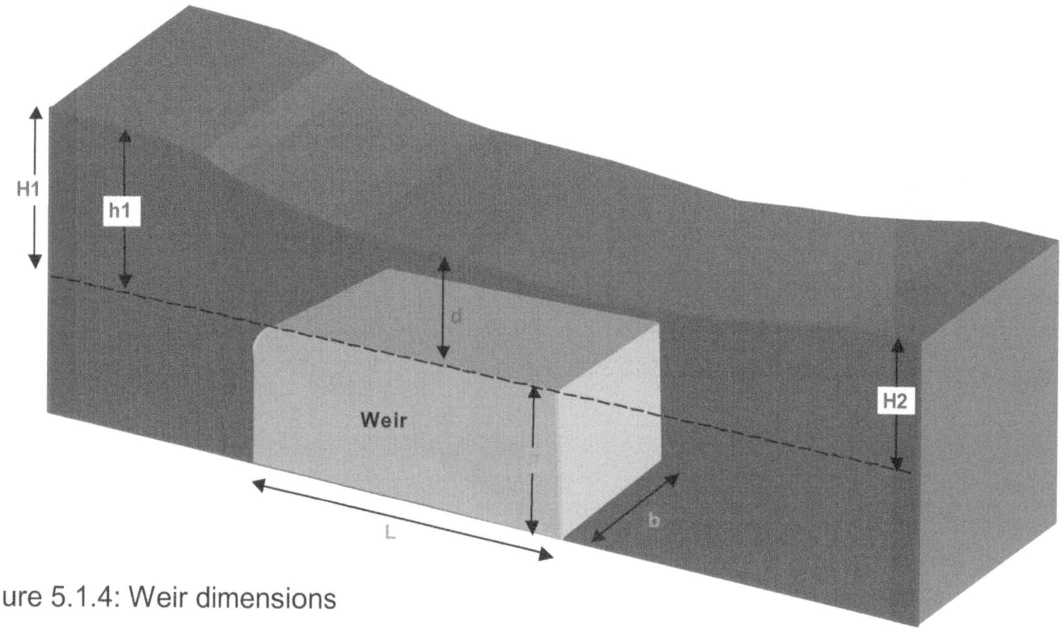

Figure 5.1.4: Weir dimensions

These design standards are taken mainly from BS 3680 (1990), although there are several other International standards. In general, the weir should occupy the full width of a lined, rectangular approach channel (Figure 5.1.4), or it should be contained between vertical abutments. The remaining dimensions are as follows. Note that H(T) is the design head, thus :

$$H(T) = H_{1max} + P$$

1. The maximum value for H_1/P is 1.5. This maximum gives a minimum value for P, hence it is the most economical and has been used for this project. When combined with $H(T) = H_1 + P$, then $P = 0.4 * H(T)$.
2. The weir length, $L \geq 1.75 * H_{1max}$, and thus for the best economy $L = 1.75 * H_{1max}$. This reduces to $L = 1.75 * 0.6H(T) = 1.05H(T)$.
3. The weir breadth, $b \geq 0.30$ m, therefore a minimum value of b = 1 m is chosen for this project. Additionally, $b \geq H_{1max} \geq 0.6 * H(T)$, and this is coded as follows:

```
If SpinButton2.Value < 0.6 * HT Then SpinButton2.Value = Int(0.6 * HT) + 1
```

4. The weir height, $P \geq 0.15$ m, thus a minimum value of H(T) = 0.5 m is chosen for this project.
5. The height of the weir must be adjusted so that the weir operates within the modular range throughout. The modular range chosen for this project is H_1/P = 0.1 to 1.5, and design, modular curves (Ackers et al, 1978) give:

$$H_2/H_1 = 0.0726 \ln(H_1/P) + 0.8637$$

Thus H_2/P can be calculated given H_1/P, and the appropriate design made. (Note that the latter may involve changing the maximum value of H_1/P; it is neglected herein.)

Boundary roughness at the weir crest

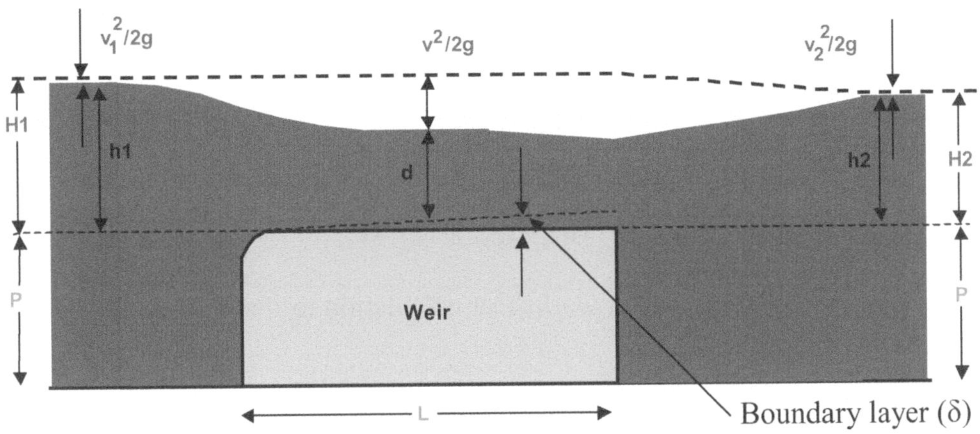

Figure 5.1.5: Side elevation of weir

The weir crest behaves as a flat plate with a developing boundary layer due to the rough surface (Figure 5.1.5). Within the boundary, the flow decellerates from the free stream velocity to zero. The boundary layer thickness (δ) is well researched for a flat plate, and summarised in Ackers et al (1978). When the breadth of the weir is much larger than the depth of flow (d), the unit discharge (flow per unit breadth) becomes:

$$q = 2/3(2/3*g)^{0.5}(H_1 - \delta)^{1.5} = 2/3(2/3*g)^{0.5} * C_d H_1^{\ 1.5}$$

where C_d is called a drag coefficient and depends, amongst other variables, on the estimated crest roughness (k). The functional form of C_d is given by:

$$C_d = (1 - 2\delta/L*L/b)(1 - \delta/L*L/H_1)^{1.5}$$

Although δ/L can be evaluated for both laminar and turbulent flow, it is unlikely that laminar flow will occur for this field project design. (Laminar flow only occurs for slow flows in depths of a few millimetres.) A turbulent rough design equation was thus derived as:

$$\delta/L = 0.0445 * L/k^{\ 0.2743}$$

and used for the evaluation of C_d. The total discharge, Q, is finally estimated as:

$$Q = 2/3(2/3*g)^{0.5} * bC_d H_1^{\ 1.5}$$

Velocity coefficient

The assumption of neglecting the approach velocity, v_1, has been made throughout this discussion. However, BS 3680 (1990) recommends that a head measurement (h_1 say) can be made from 1 to 4 times H_{1max} upstream of the weir nose (Figure 5.1.5). It is then possible to correct for v_1 by an iterative computation. The stage-discharge equation reduces to:

Equation 1: $$Q = C_d * g^{0.5}\, bH_1^{1.5}$$

Since $Q = A_1v_1$, where $A_1 = b(h_1 + P)$ is the cross section area at the measurement section, then H_1 is given by:

Equation 2: $$H_1 = h_1 + Q^2/2gA_1^2$$

These 2 equations may be used in a worksheet calculation to iterate to a true solution for Q as follows:

1. For the initial h_1 value, estimate A_1.
2. Using h_1 in place of H_1 in Equation 1, estimate Q.
3. Estimate a new value for H_1 using Equation 2 and the Q value.
4. Recalculate Q in Equation 1 using the H_1 value fom the previous step.
5. Repeat steps 4 and 5 to convergence.

As an alternative to iteration, a velocity coefficient may be used. The expression for Q then becomes:

$$Q = C_vC_d * g^{0.5}\, bH_1^{1.5}$$

The coefficient, C_v is given implicitly by:

$$C_v^2 = (1 - 0.5R_aC_v^2)^3$$

where: $$R_a = 0.544C_dbh_1/A_1$$

The variable, C_v, may however be solved explicitly by the following polynomial:

$$C_v = 955.53\, R_a^6 - 5779.1\, R_a^5 + 13954\, R_a^4 - 16815\, R_a^3 + 10115\, R_a^2 - 2429.9\, R_a$$

5.1.4 Modules

There are 2 modules used for this code. The first computes the design, operation and flow discharge, whilst the second manipulates the moving water surface in the "Elevation" drawing. They are described in sequence.

"Weir"

The 3 subroutines in this module are straightforward calculations of the above equations. The "Design" subroutine reads H(T) then calculates P and L . The breadth b is also read, and adjusted for a minimum value as described above. The "Operation" subroutine reads the H_1/P input, then calculates the d/P and H_2/P ratios. And finally, the "Discharge" subroutine reads the roughness value selected from the "Run" worksheet using a logical If condition. The Cd value is then calculated and used to estimate the flow discharge.

"Levels"

There are 3 subroutines used in this module. The water level surface is divided into 3 sections of equal depth (H1, H2 and d), an upstream section, and a differing downstream section. Each subroutine is complicated by the fact that water levels must be described in worksheet plottting units. As a reminder, these units have their origin at the top left hand corner of the worksheet. A worksheet cell with a default dimension of 80 pixels (width) by 22 pixels (depth) has plotting dimensions of 48 by 13.2 units.

The "EqualDepths" subroutine plots the changing surface and energy lines in 5 steps, as follows:

1. The H1 shape and dimension line are plotted. Note that the latter assumes that $v_1^2/2g$ is zero, and is extended to the total energy line. The H1 shape is simply changed by adjusting the top level.
2. The critical depth shape and dimension line are plotted.
3. The H2 shape and dimension line are plotted.
4. The total energy line is plotted. Note that there is a fall in total energy over the hydraulic jump, and that energy lost to fluid friction at the bed is neglected.
5. The velocity-head dimension line for d is plotted. This plot becomes complex if H_1/P is reduced below 0.3, as the editing of both points on the arrow indicator line do not allow the arrowheads to move. The dimension line is therefore excluded for small lengths by using a logical If condition.

The "Upstream" subroutine plots the upstream levels by using the H_1/P values as a factor for the original highest level coordinates. The "Downstream" subroutine is coded in a similar way, except that the coordinates nearest to H_2 are factored according to H_2/P. The final result is to approximately simulate a changing water surface for modular flow.

5.2 FLOOD FREQUENCY USING ANNUAL MAXIMUM FLOWS

5.2.1 Summary

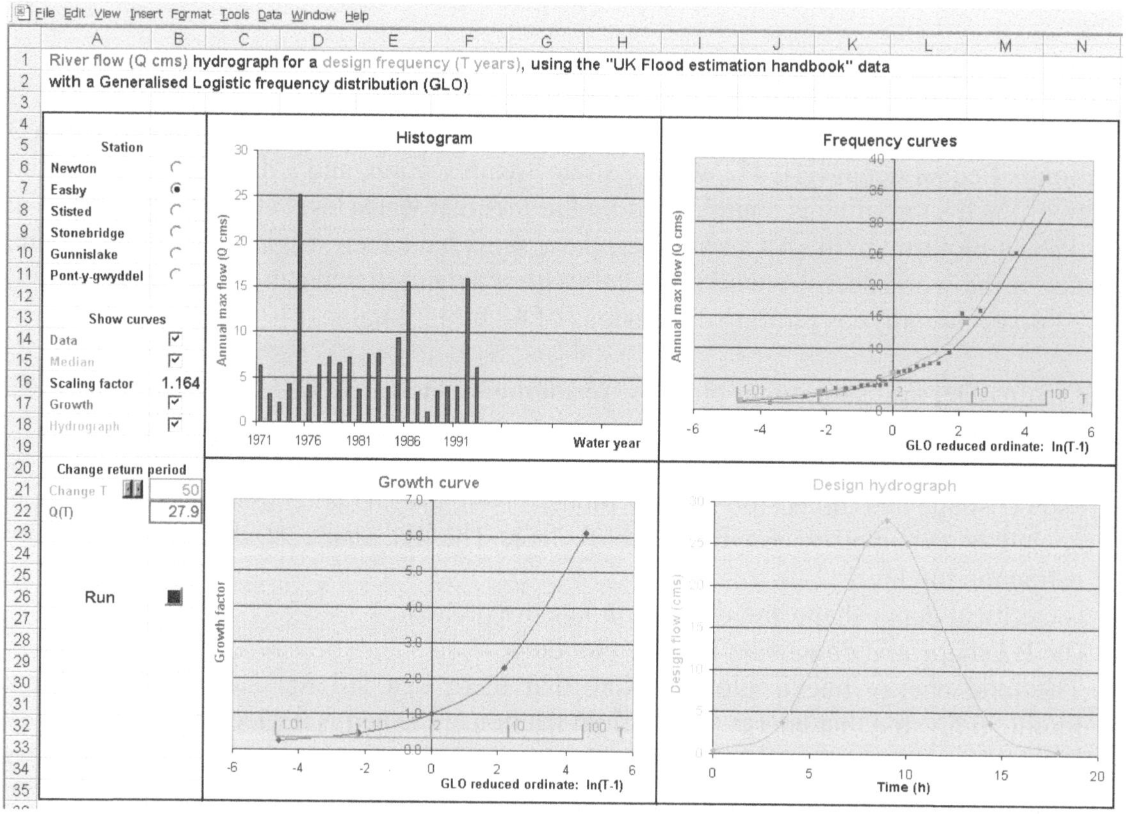

Figure 5.2.1: The "Run" worksheet for the "QAM.xls" project

This project uses the "Flood Estimation Handbook" method (Volume 3, Robson and Reed, 1999) for 6 river flow gauging stations to estimate the frequency of occurrence for extreme floods, and plot a design hydrograph (Figure 5.2.1). The data are abstracted from the FEH data bank of about 1000 stations, and each of the 6 records contains about 20 years of data. In accord with the FEH recommendations, the flood frequency distributions are calculated using annual maximum flows (the largest flood discharge in a given year, Q m^3/s) when the record length exceeds 14 years.

The flood frequency curves are derived by multiplying a "growth curve" distribution and an "index flood" value. The latter is chosen as the middle value of the data record (the median, QMED). Since half of the data values are larger than QMED and half are less, the probability of occurrence (P) for QMED is 0.5, and its recurrence interval (return period, T = 1/P) is 2 years. In cases where a data record is an even number, QMED is taken as the mean of the 2 numbers in the middle of the set. The growth curve uses the Generalised Logistic frequency distribution (GLO) as detailed previously in the "LMoments.xls" project. This distribution is first fitted to the actual data, then again to a normalisation of the data (the growth curve). The normalisation is made such that the ordinate (growth factor) is unity when the abcissa is T = 2 years.

5.2.2 Code operation

The project is operated by selecting a gauging station using an Option button, then activating the "Run" Command button. The Option button loads the relevant data from the "Data" module and a new histogram is plotted on the "Run" worksheet. The Command button runs all the subroutines in the "GLO" module, and completes the remaining 3 charts. In addition, a scaling factor (described below) and the flood peak discharge are computed and displayed on the "Run" worksheet.

A return period (T) is selected using the Spin button, and its scrolling value properties are set from 2 to 100 years. The latter may be changed at will, and a new computation is made on reactivating the "Run" Command button.

The 4 Check boxes for each of the 4 curves are present to identify the curves. When the Check boxes are ticked, the curves are visible and vice-versa. This is achieved by the Private control subroutines in the Sheet1 (Run) object. The subroutines use a logical If condition to determine whether a Check box is ticked (true), and set to the appropriate colour of the curve and data marker. Another If condition is used in the subroutines to determine whether the "Run" command has been activated (by checking the "scaling factor" for data). If the "Run" command has not been activated, a Message box warning is used. (The latter is necessary as results from the previous run will occur.)

Station	Newton	Easby	Stisted	Stonebridge	Gunnislake	P.Gwyddel
T years	2	2	2	2	2	2
Q m^3/s	6.9	6.1	13.3	29.5	264.9	63.2
T years	50	50	50	50	50	50
Q m^3/s	16.3	27.9	18.8	58.0	576.4	145.0
T years	100	100	100	100	100	100
Q m^3/s	19.4	37.2	19.3	64.5	670.1	171.3
Scaling factor	0.992	1.164	1.028	1.006	0.947	0.964

Table 5.2.1: Some typical results for flood peak values

Table 5.2.1 summarises some typical flood peak values (Q m^3/s) for selected return periods (T years). Note how the Stisted T = 100 year magnitude is close to the T = 50 year value. This is due to the curve being "upper bounded", as described below.

The stations used for this project are the same as those chosen as examples in FEH, Volume 3. The results may be compared, and found to be the same. The Hosking and Wallis (1997) method of fitting L-moments has been adopted, in sequence to the same method used in the "LMoments.xls" project. It is therefore recommended that the latter project be studied first, as it is frequently referenced herein to avoid repetition.

5.2.3 Procedure

Frequency distributions

There are 3 frequency distributions associated with this project, namely the FEH data distribution, the growth distribution and the median distribution. The inverse distribution function (Q(F)) for the 3 parameter GLO distribution is first fitted to the FEH data, thus:

GLO 3 parameter curve: $Q(F) = \xi + (\alpha / k) [1 - ((1- F) / F)^k]$

where: $\xi = l_1 - \alpha [1/k - \pi / \sin k\pi)]$; $\alpha = l_2 \sin k\pi / k\pi$; $k = t_3$

The L-moment ratios are described in the "LMoment.xls" project. This data distribution is normalised with the F = 0.5 (Q(F) = ξ) value to give the growth curve. The ordinate is named the growth factor (x(F)), and is thus given by:

growth factor: $x(F) = Q(F)/ \xi = 1 + (\alpha / \xi k) [1 - ((1- F) / F)^k]$

Finally, the median distribution is obtained by multiplying the median data value (QMED) by the growth curve, thus:

median curve: $Q(F)_{MED} = QMED * x(F)$

Note that the median curve is not the same as the data curve. It differs by a constant scaling factor (QMED/ ξ) due to the fact that L-Moments are derived using the average values of the data. As noted by FEH however, there are only 11 out of 421 stations with long records (> 20 years) where the scaling factor > 1.10.

All three curves are plotted against the logarithm of the GLO ordinate, -ln((1-F)/F). As with the Gumbel reduced ordinate, the use of a logarithmic scale moves the data towards a straight line. For these project examples, all of the median curves are lower bounded (convex downwards, for which k < 0) except the Stisted station (upper bounded, k > 0).

Design hydrograph

For many design applications such as the "ResRoute.xls" project below, a design hydrograph is required using the relevant peak flood magnitude. Since all hydrographs differ, the FEH recommends the choice of 3 optional methods. The method of borrowing a standard hydrograph shape from the rainfall-runoff analysis (described in the "FEH.xls" project) is illustrated herein. And since the stations chosen are all from rural catchments, the FEH "75% winter profile" is used as opposed to the "50% summer profile" designed for urbanised catchments.

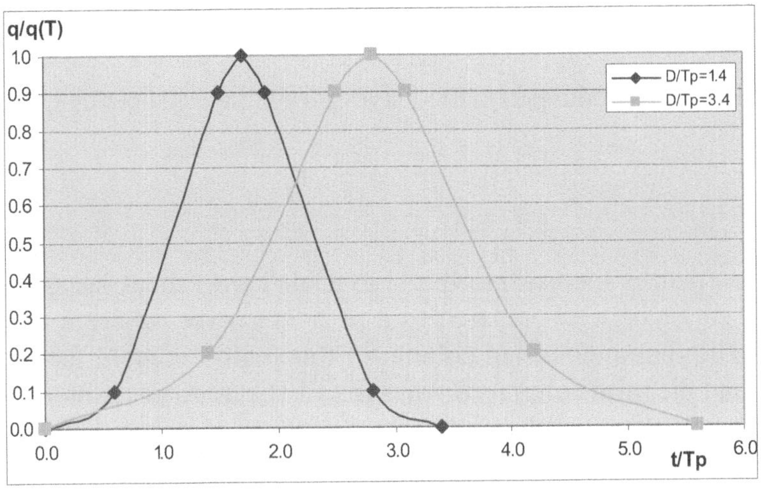

Figure 5.2.2: Standard hydrograph shape for a "75% winter profile"

The standard hydrograph shape is determined by an indicator variable, D/Tp (where D is the flood duration, and Tp the time taken for the unit hydrograph to reach peak flow). This indicator may be estimated from FEH, Equation 4.3.1, as:

$$D/Tp = 1 + SAAR/1000$$

where SAAR is the standard average annual rainfall for the stations in mm. A low value for SAAR is 400 mm and a high value is 2,400 mm. Thus the variable D/Tp ranges from 1.4 to 3.4, and the standard hydrograph shapes appear as in Figure 5.2.2 (approximated from FEH, Figure 4.3.9).

The standard hydrograph shape is almost symmetrical, and may be approximated by 7 coordinates. The D/Tp = 1.4 hydrograph is approximated to:

t/Tp	0	0.6	1.5	1.7	1.9	2.8	3.4
q/q(T)	0	0.1	0.9	1.0	0.9	0.1	0

where t is the time in hours, q is the instantaneous flow in m^3/s and q(T) is the peak flow for a T year return period. Similarly, the D/Tp = 3.4 hydrograph is approximated to:

t/Tp	0	1.4	2.5	2.8	3.1	4.2	5.6
q/q(T)	0	0.2	0.9	1.0	0.9	0.2	0

The standard hydrograph may therefore be synthesised by linear extrapolation from the D/Tp = 1.4 hydrograph, given the D/Tp value for any station. The detail of this extrapolation is contained in the "Hydrograph" worksheet, which uses worksheet formulae.

5.2.4 Modules

This project consists of 3 modules and the "Hydrograph" worksheet calculation. They are now described.

"Data"

Each set of station data is loaded into the "FEHData" worksheet using the "LoadData" subroutines. The data has been prepared as a (n * 3) matrix, where n is the number of years of record. The first and second column of this matrix contain the year and annual maximum flow, and the third column contains the FEH station descriptors which are used in the hydrograph calculation. When alternative data are to be used, these matrices may simply be replaced.

The annual maximum flow data is copied to the "Design" worksheet and the station descriptors to the "Hydrograph" worksheet, using the "Copydata" subroutine. In addition, the rank, Gringorten probability, return period and GLO ordinate are computed on the "Design" worksheet.

"GLO"

The GLO module begins with the "LMomentsAA" and "GLOQuantiles" subroutines relevant to the original data. These subroutines are similar to those described and used in the "LMoments.xls" project. The module continues with similar subroutines for the "Growth" curve, namely "LmomentsGG" and "NGLOQuantiles". The new data used in column G of the "Design" worksheet is however the original data divided by the L–Moment parameter ξ (a Greek symbol pronounced "Ex"). Both sets of L-Moment ratios are written to the "Design" worksheet, and it is seen that l1 and l2 for the "Growth curve" differ from the "Data" curve, as expected by definition.

The "NGLOQuantiles" subroutine also computes the median frequency data for the 5 quantiles, simply by multiplying the growth factors by QMED. The 3 sets of quantiles (for the data, growth and median) are written to the "Design" worksheet for plotting in the relevant "Run" worksheet charts. Finally, the last subroutine in the module, "SolveQ", computes the Q value for any T value from 2 years to 100 years (using the L–Moments of the growth curve) and enters it on the "Run" worksheet.

"Charts"

There are 4 line-chart series in this project, namely the data and median frequency curves, the growth curve and the design hydrograph. The first subroutine in this module, "EditCharts", makes these curves invisible when a new station Option button is selected. This is achieved by setting the chart lines, borders and markers to xlNone. Each curve is then made visible when the "Run" Command button is activated. This is done using the subroutines "StationData", "Median", "Growth" and "PlotHydrograph".

"Hydrograph" worksheet

All calculations for the design hydrograph are made using worksheet formulae, to illustrate this alternative method. The calculations follow the enumerated FEH method, and are in 7 steps, as follows:

1. Baseflow (BF) :- the total BF for the catchment area is evaluated from the standard average annual rainfall (SAAR) using the condition defined in the "FEH.xls" project:

If SAAR < 800 Then $BF = 10^{-2} * Area\ (0.008SAAR)$
If SAAR > 800 Then $BF = 10^{-2} * Area\ (0.003SAAR)$

2. Response (q) :- the BF is subtracted from the computed peak flow (Q(T)), thus:

$$q = Q(T) - BF$$

3. D/Tp indicator :- where D is the storm duration is estimated using the SAAR, thus:

$$D/Tp = 1 + SAAR/1000$$

4. Tp :- the unit hydrograph time to peak is estimated using the catchment parameters, thus:

$$Tp = 1.1\ [4.270 DPSBAR^{-0.35}\ PROPWET^{-0.89}\ DPLBAR^{0.54}\ (1 + URBEXT)^{-5.77}]$$

5. The dimensionless ordinates for the standard design hydrograph are computed by extrapolation in Cells D23:E29.

6. The t/Tp ordinate is multiplied by Tp in Cells A36:H37.

7. The q/Q(T) ordinate is multiplied by Q(T) in Cells A38:H39 and the baseflow is added in Cells A40:H41 to give the final design hydrograph. The t and q ordinates are then used in the "Run" worksheet for the design hydrograph chart.

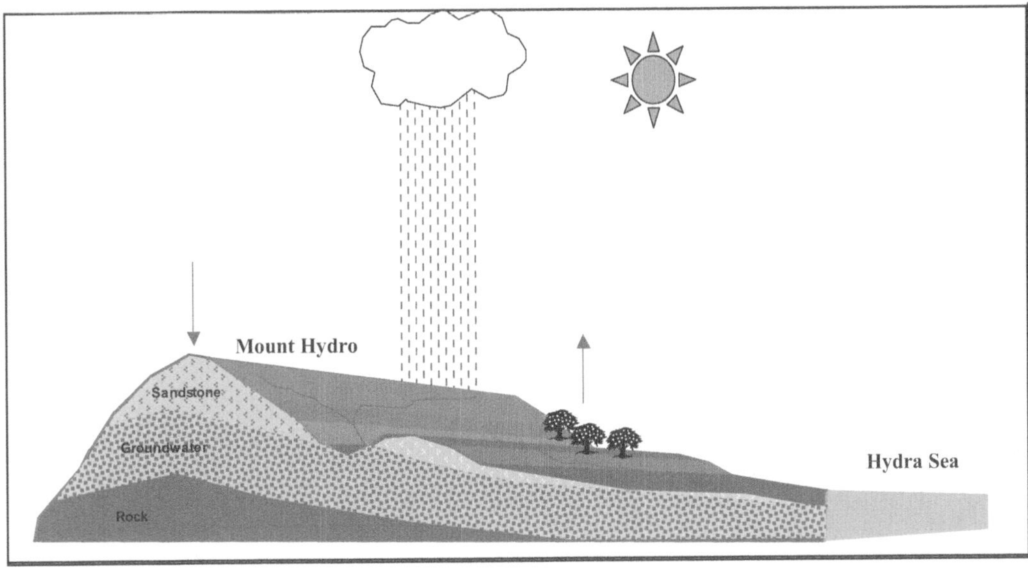

Hydrologic cycle

6.1 RAINFALL-RUNOFF USING THE FEH METHOD

6.1.1 Summary

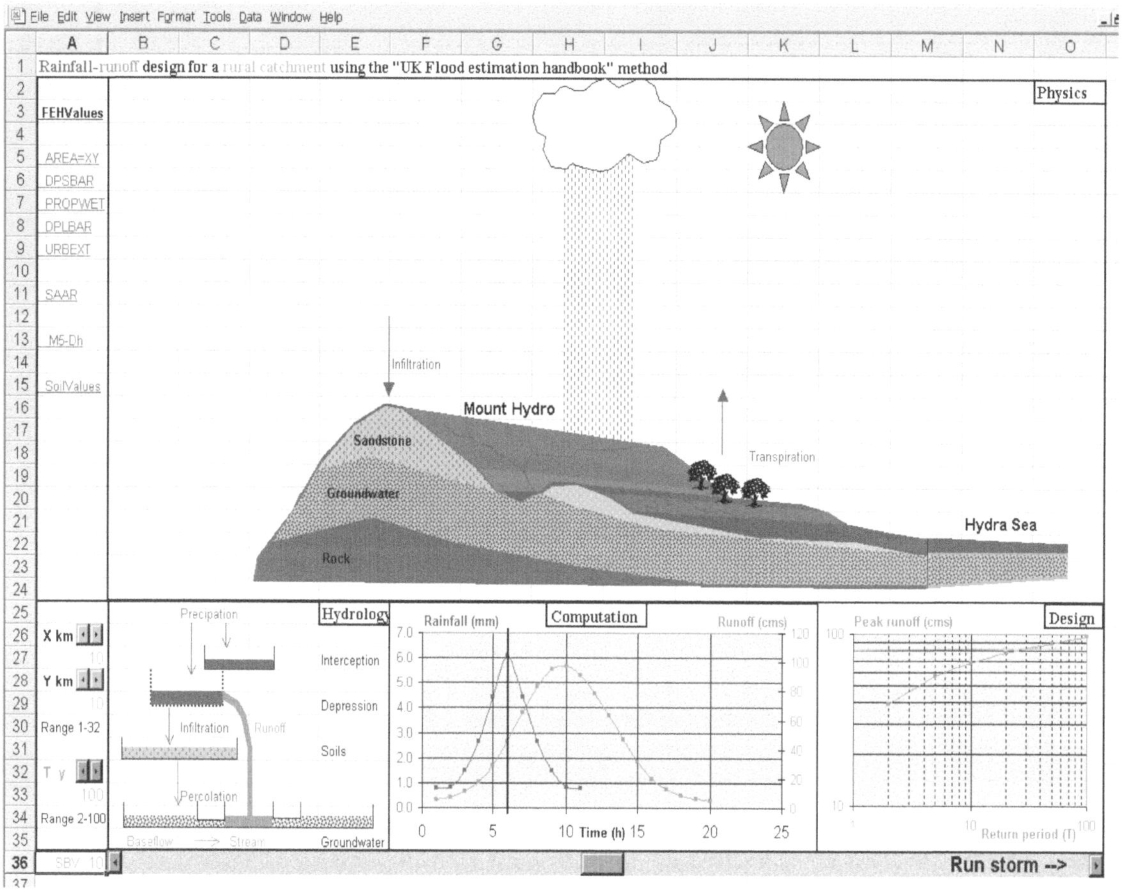

Figure 6.1.1: The "Run" worksheet for the "FEH.xls" project

The ultimate challenge for hydrological engineering is to predict a design river flow for a known incident rainfall to a catchment. In most countries, there are adequate records of rainfall, but fewer records of the more difficult and expensive measurement of stream flow. The prediction of stream flow requires a detailed knowledge of the entire hydrological cycle in both space and time. This involves at least, the physical, hydrological, computational and design aspects of the process.

The basic principle for rainfall-runoff modelling is to find the volumetric loss of water mass due to such processes as evaporation, transpiration, interception, depression storage, soil moisture storage and groundwater recharge. Many detailed deterministic (event based) and stochastic (continuous simulation) models exist. The FSR unit hydrograph and losses model (FEH, Volume 4, 1999) is used herein since it can be easily generated for any site and permits the inclusion of local data (Figure 6.1.1).

6.1.2 Code operation

The Hydrological Cycle

The moving objects of the "Run" worksheet show the major processes of the hydrological cycle. The physical area illustrates the passage of a storm cloud over a mountainous area, from land to sea. And the hydrological area attempts to define the main processes throughout the event, and render them available for mathematical analysis.

Precipitation as snow, hail and rain occurs when the moist air in clouds condenses as it is lifted and cooled. In many global areas, winds are generated from a prevailing direction, and this orographic rainfall causes a prevailing rainfall trend. There are other climatic conditions that cause precipitation, such as the convergence of air masses at different temperature (frontal rainfall) or the terrestial heating (convective rainfall) or cooling (mist and fog) of moist air masses.

When precipitation reaches the ground, part of it is intercepted by the vegetation surface, and eventually evaporates with associated transpiration. Another part is collected in depressions such as lakes, and is stored for later participation in the cycle. The water surface for the depression storage is again subject to evaporation.

Both gravity and molecular forces cause the precipitation at the ground to infiltrate into surface soils. Deeper soils are generally more compact, and a saturation zone develops at a deep level. The groundwater in this zone collects, and moves with the hydraulic gradient as baseflow.

The mass of precipitation in excess of the loss processes of evaporation, transpiration, interception, depression storage and infiltration contribute to surface runoff. The latter concentrates by gravity into streams and rivers, where it combines with groundwater baseflows to produce a total streamflow.

A brief summary of these events is that although both rainfall and runoff are measurable as surface quantities, it is the loss processes that are mainly hidden and therefore difficult to quantify and theorise. The FEH method used herein uses the empirical method of applying data from actual streamflow measurements. An archive of about 3360 flood events has been compiled for about 250 UK streamflow stations (Houghton-Carr and Boorman, 1991). This is used to define the major parameters that influence the precipitation losses (for example, the Soil Moisture Deficit). A triangular "Unit Hydrograph" (UH) is generated to define the streamflow that would occur for a unit value (say 10 mm) of rainfall (in excess of baseflow) over a period of unit time (say 1 hour). The UH is then convoluted for the actual design storm magnitude and duration. Finally, the baseflow is added to give the total streamflow.

Code events

There are 2 main Controls that operate this project. The first is the Scroll Bar at the top of the "Run" worksheet; it runs and restarts the entire storm. The Scroll Bar Value (SBV) varies in integers from 0 to 20 and determines the events that occur visually in the "Physics" ("Pbox"), "Hydrology" ("Hbox") and "Computation" ("Cbox") areas of the worksheet (Table 6.1.1). Note that the storm only runs forward with the Scroll Bar control. An additional Control required to reset the storm at any SBV is left as an exercise. The second main Control is the "Return Period" (T) Spin button that runs the Extreme Value code for the "Design" chart of the worksheet.

The 2 remaining Spin buttons (X and Y) control the catchment area by varying the catchment length and width. Their values are abritrarily set from 1 to 32, so that a maximum catchment area of about 1000 km^2 may be modelled. The other FEH hydrological variables are adjusted in the "Design" worksheet. They may be accessed from the "Run" worksheet by use of the [Hyperlinks].

Area Event Shape	Pbox Cloud FF 42	Pbox Rain FF 37	Pbox Lightning AS 52-54	Pbox Lake level FF 7	Pbox GW level FF 43	Pbox Arrows FF 31,34	Pbox Lighting FF 2	Hbox Water levels FF 111-115,123	Hbox Runoff FF 89	Hbox Arrows FF 100-104	Cbox Cursor FF 118
SBV 0	Start										
SBV 1	Up										
SBV 2	Up										
SBV 3	Up										
SBV 4	Up										
SBV 5	Up										
SBV 6	Up	Light				Infiltration	Normal		Draw	Infiltration	Draw
SBV 7	Up	Light		Increase	Increase			Increase	Increase		Move right
SBV 8	Up	Heavy	AS 52 on	Increase	Increase			Increase	Increase		Move right
SBV 9	Up	Heavy	AS 52 off	Increase	Increase			Increase	Increase		Move right
SBV 10	Up	Heavy	AS 53 on	Increase	Increase			Increase	Increase		Move right
SBV 11	Down	Heavy	AS 53 off	Increase	Increase			Increase	Increase		Move right
SBV 12	Down	Heavy	AS 54 on	Increase	Increase	Interception		Increase	Increase		Move right
SBV 13	Down	Heavy	AS 54 off	Increase	Increase		Dim	Increase	Increase		Move right
SBV 14	Down	Light		Increase	Increase			Increase	Increase		Move right
SBV 15	Down	Light		Decrease	Decrease		Normal	Decrease	Decrease		Move right
SBV 16	Down			Decrease	Decrease	Evaporation	Bright	Decrease	Decrease	Evaporation	Move right
SBV 17	Down			Decrease	Decrease			Decrease	Decrease		Move right
SBV 18	Down			Decrease	Decrease			Decrease	Decrease		Move right
SBV 19	Down			Decrease	Decrease			Decrease	Decrease		Move right
SBV 20	Reset	Reset		Redraw	Redraw			Redraw	Invisible		Invisible

Table 6.1.1: Shape events controlled by the Scroll Bar Value (SBV)
(FF = Freeform, AS = Autoshape)

The shape events that move with the SBV are called in the Sheet1 (Run) object by the "ScrollBar1" subroutine. There are 3 types of operation for the 26 shapes, as follows:

1. The relative movements for some shapes are called from procedures in the "Shapes" module. These shapes are the Cloud, Lake level, Groundwater level, Hbox levels and Cursor used in the Cbox.
2. The absolute coordinates for the "Hbox" Runoff shape is coded, for every change, in the Private subroutine.
3. All other shape events are called in the Private subroutine.

6.1.3 Procedure

Although the FEH method is fully documented and physically described, several small changes have been made herein to allow a simplified and continuous computer simulation for the entire method. For example, the simulation is for a rural catchment only, the Tables and Figures referenced in the FEH text have been converted to functional equations and the UH time is restricted to 1 hour. The method is reproduced in the following 8 steps, and these are directly coded as worksheet functions in the "Design" sheet.

1. Unit Hydrograph

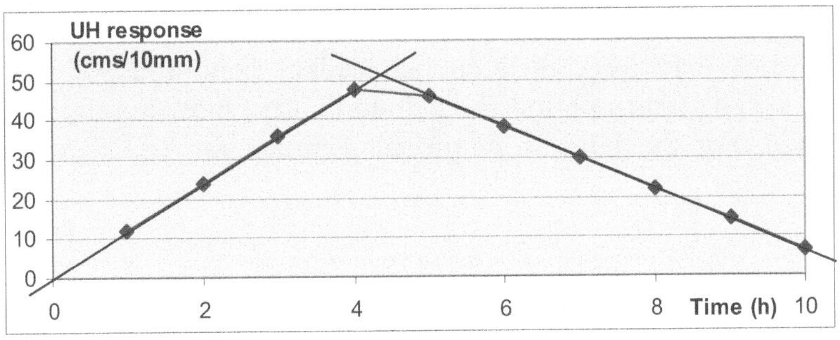

Figure 6.1.2: FEH triangular unit hydrograph

Four variables are used to define a triangular UH, in addition to the catchment area (Figure 6.1.2). These are the mean drainage slope (DPSBAR), the proportion of time that the soil moisture deficit was below 6 mm during the 1961-1990 period (PROPWET), the mean drainage distance (DPLBAR) and the extent of urban and supurban land cover (URBEXT). The time to peak flow of a UH (Tp in hours) is estimated from the time to peak for an instantaneous UH (Tp(0)), as the latter is simpler to determine, thus:

$$Tp(0) = 4.270\ DPSBAR^{-0.35}\ PROPWET^{-0.80}\ DPLBAR^{0.54}\ (1 + URBEXT)^{-5.77}$$

Since we are considering an average value for the 1-hour rainfall duration, then:

$$Tp = Tp(0) + ½$$

Note that if a site is gauged, it is better to estimate Tp from the actual data or from an observed estimate of the time between rainfall and runoff peaks.

The UH peak flow (Up) and time base (TB) are defined empirically as:

$$Up = AREA * 2.2/Tp \quad \text{and} \quad TB = 2.52\ Tp$$

The UH in Figure 6.1.2 uses example data for a rural UK catchment with AREA = 100 km^2. The coordinates are computed by the "UnitHydrograph" subroutine in the "Hydrology" module.

2. Design storm duration

The storm duration (D) is mainly a function of the catchment response (represented by Tp). A secondary influence due to the overall climate has however been noted, and this is represented by a standard, annual average rainfall variable (SAAR), thus:

$$D = Tp\ (1 + SAAR / 1000)$$

Since the design storm profile is assumed to be symmetric, D is rounded to the nearest odd integer.

3. Design storm depth

Similar runoff events may have different antecedant conditions of storm depth and catchment wetness. The return period for a flood (T say) may therefore differ from that of the rainfall (T_R). For the UK, an empirical estimate for T_R is evaluated from the recommended curves as:

$T \leq 30$ y: $\quad T = -0.011T_R^{\ 2} + 2.0919T_R - 2.8141$
$T > 30$ y: $\quad T = 2.8351T_R^{\ 0.8492}$

The design storm depth (P) can now be found from DDF curves (as described in the "DDF.xls" project), since D and T have been estimated. To simplify the calculation, an estimate of P for T = 5 years may be used (called the 5-year D-hour point rainfall, M5-Dh), and a growth factor applied for alternative T values (to give the MT-Dh point rainfall). The growth factor is a function of rainfall return period (T_R) and storm depth (M5). A quadratic trend surface with a "goodness of fit" indicator of 0.93 was fitted to the tabulated values, thus:

$$\text{Growth Factor} = 0.368 + 0.736\text{Log}(T_R) + 0.323\text{Log}(M5) + 0.00921(\text{Log}(T_R))^2 - 0.0797(\text{Log}(M5))^2 - 0.141\text{Log}(T_R)\text{Log}(M5)$$

This surface does not however extend to the 10,000 year return period, since the tabulated data were incomplete.

Finally, there remains the conversion of this point rainfall depth to a rainfall depth for the catchment area. It is inevitable that the latter will be less, and curves of Areal Reduction Factor (ARF) that vary with AREA and D have been used in the past (Stewart, 1989). These ARF curves can be approximated to within a few percent by evaluating the function:

$$S = (\text{Log}(AREA) + 10) / (\text{Log}(D) + 8)$$

And solving for ARF using:

$$ARF = 0.01\ (-40.534S^2 + 55.938S + 86.512)$$

4. Design storm profile

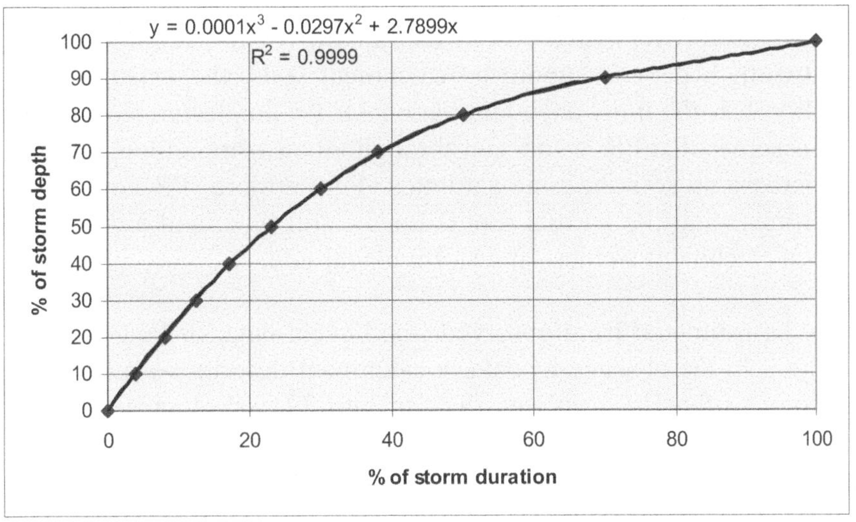

Figure 6.1.3: Design winter storm profile as a cumulative percentage related to storm peak

The storm profile is used to distribute the design storm depth (P) within the design storm duration (D). For a rural catchment, the 75% winter profile is recommended for design. This is defined to be, on average, of greater kurtosis (peakedness) than 75% of the UK winter storms. The profile is symmetrical and bell-shaped, having a cumulative percentage of depth against discharge as shown in Figure 6.1.3. Note that the percentage of the storm duration is measured from the storm peak outwards, thus ensuring that the storm profile is symmetric. The cumulative distribution may be approximated by a cubic polynomial to within 1 percent, thus:

$$\%P = 0.000118(\%D)^3 - 0.0297(\%D)^2 + 2.7899(\%D)$$

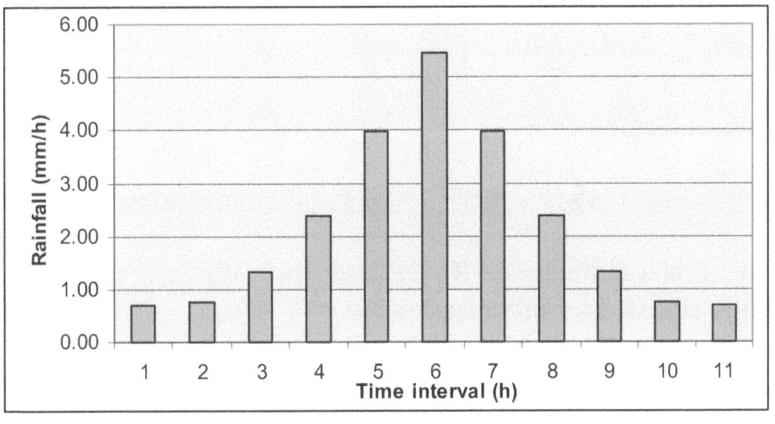

Figure 6.1.4: Design winter storm profile

The "StormProfile" subroutine in the "Hydrology" module computes the storm profile for the example herein (Figure 6.1.4). It is described later, and may be used for any duration.

5. Standard percentage runoff

The percentage runoff (PR) is that part of the rainfall which becomes the direct response runoff to a stream. It depends upon the catchment soils, the existing wetness and the storm variability. It is the most uncertain parameter for the design storm hyetograph. The FEH method assumes that PR is constant throughout a storm, although it is fully realised that it will increase as soils become saturated. Furthermore, PR consists of a standard term (SPR) representing the normal runoff for the soil type, and a dynamic term (DPR) dependent upon catchment wetness and storm magnitude.

The SPR may be estimated from observed flood event data, or long-term measurements of the base flow. In the absence of data, a catchment descriptive method must be used. The soil geology of the UK has been divided into 29 hydrological types called HOST classes. Each class is in turn associated with a particular SPR. A chalk soil will have a low SPR of about 2% and a clay soil SPR will be high at 60% say. A design SPR for a heterogeneous catchment may thus be estimated as :

$$SPR = {}_i \sum SPR_i \ HOST_i$$

6. Dynamic percentage runoff

The wetness state of the catchment prior to a storm is estimated by a Catchment Wetness Index (CWI). A design CWI is a function of the standard annual average rainfall, and is given:

for SAAR < 800 mm: $CWI = 0.3\,(SAAR) - 120$
for SAAR > 800 mm: $CWI = 0.0025\,(SAAR) + 121$

The DPR depending upon CWI is then:

$$DPR_{CWI} = 0.25\,(CWI - 125)$$

and that for the storm magnitude is:

$$DPR_{RAIN} = 0.45\,(P - 40)^{0.7} \quad \text{for } P > 40 \text{ mm, and is zero otherwise}$$

For a completely rural catchment SPR, DPR_{CWI} and DPR_{RAIN} are summed. The sum (PR_{RURAL}) is finally adjusted for urbanisation by:

$$PR = PR_{RURAL}\,(1 - 0.615\,(URBEXT)) + 70\,(0.615\,(URBEXT))$$

This final PR value is used in the "StormProfile" subroutine to compute the net rainfall hyetograph.

Total runoff hydrograph

The convolution of the UH and the net rainfall hyetograph is computed in the "Convolution" subroutine in the "Hydrology" module and displayed in the "Design" worksheet. The baseflow (BF) must finally be added to the hydrograph ordinates. As for the rainflow design events, baseflow may be estimated from observed flood event data. In their absence however, catchment descriptors may be used as follows:

$$BF = (33(CWI - 125) + 3.0\ (SAAR) + 5.5)\ 10^{-5}\ AREA$$

For some catchments the BF may evaluate to a negative value; it is then set to zero.

8. Flood peak frequency

This project may finally be used to demonstrate flood peak values for a range of return periods (T). The maximum flow (or flood peak) in the final design hydrograph is evaluated using the "EVCurve" subroutine in the "Hydrology" module.

The flood peak is associated with the "return period" (T) Spin button in the "Run" worksheet. When the Spin button is scrolled through its range of values (2 to 100), the "EVCurve" subroutine will record the peak values and store the 2, 5, 10, 20, 50, and 100 year peak flows in the "Design" worksheet. These ordinates are also plotted in the "Design" area of the "Run" worksheet.

6.1.4 Modules

"Shapes"

The moving shapes are the Cloud, Lake level, Groundwater level, Hbox levels and Cursor. This module programs both the movement of entire shapes (such as the cloud) or simply a shape surface (such as the lake). The procedures which move the entire shape ("MoveCloudUp", "MoveCloudDown", "MoveCursor") use code which reposition the entire shape by a defined number of worksheet drawing units, thus:

```
ActiveSheet.Shapes("Freeform 42").Select
Selection.ShapeRange.IncrementLeft 28.8
Selection.ShapeRange.IncrementTop -6.6
```

The remaining procedures scale the relevant dimensions by a multiplicative factor, thus:

```
ActiveSheet.Shapes("Freeform 7").Select
Selection.ShapeRange.ScaleHeight 1.1, msoFalse, msoScaleFromBottomRight
Selection.ShapeRange.ScaleWidth 1.04, msoFalse, msoScaleFromTopLeft
```

Since the scaling cannot be reversed exactly, the shapes are redrawn completely when the Scroll Bar Value is 20 ("ReDrawLake", "RedrawHBox").

"Hydrology"

Since all of the procedures in this module read and write data using the "Design" worksheet, they all begin by selecting this sheet and end by reselecting the "Run" worksheet. This sheet change could be avoided in the code by prefixing all the VB Range or Cells commands with the Sheets("Design") object. However, the code becomes less easy to read. An advantage for using the extra Sheets("Design") object is however that the return period Spin button need not be continuously depressed for activating code, and this principle is used in other projects. A summary of each procedure now follows.

"UnitHydrograph"

This subroutine calculates the UH ordinates for hourly intervals in a single loop. The code is readily understood by considering it in variable form:

```
t = 1 to Integer(TB)
If t < Tp then U = t * Up/Tp
Else U = (TB – t) / (TB – Tp)
```

"StormProfile"

Three steps are used to calculate the storm profile, as follows:

1. Define peak time of the storm profile and the time series in hours.
Since the profile is symmetric and the duration has been set to an odd integer, the peak time (DM) is given by $D_{odd}/2 = 0.5$. The time series uses the Excel commands [Edit, Fill, Series, Row] to enumerate the integer series from 1 to D_{odd}.

2. Calculate the rainfall value in the hyetograph peak hour, as follows:
   (a) $\%D = 100 / D_{odd}$
   (b) $\%P = 0.000118(\%D)^3 - 0.0297(\%D)^2 + 2.7899(\%D)$
   (c) Copy the above value to the next row for summing the %P values
   (d) Peak hour storm depth = (%P / 100) * P
   (e) Copy the peak hour storm depth into the middle column of the hyetograph

3. Calculate the storm depths for the upper and lower hours of the hyetograph.
The depths are calculated in a similar sequence to the peak hour, using a loop of (DM – 1) values for the equal depths of the upper and lower hours. Note that the %P value from a preceeding hour must be subtracted from a current hour since values are taken from a cumulative curve. The data in (c) and (d) are also summed to check for correctness. Finally, the percentage runoff value (PR) is applied to give the rapid response hyetograph.

"Convolution"

This subroutine uses 4 steps as follows:

1. Delete previous flood hydrograph thus ensuring there is no corruption of data.
2. Write storm profile in a column from Cell A60 downwards (mm/10). Note that the vertical looping variable uses the storm duration.
3. Compute the hydrograph abcissa values for each duration time interval. Note that the vertical looping variable uses the storm duration, and the horizontal (abcissa) loop uses the timebase.
4. Displace each abcissa row to right using the vertical looping variable.

"SumColumns"

This subroutine begins by calculating the constant baseflow, then adds it to each of the summed hydrograph columns. It ends by writing the hourly times for each column.

"EVCurves"

The hydrograph flood peak values are found using the worksheet function [max], and each associated return period is determined by a logical If statement.

Alternative Data

In addition to the HOST soil classes defined by the FEH method, there are 7 other major variables that may change for a particular catchment. The variables of HOST, AREA and M5 – Dh can be changed at will, since they do not alter the UH timebase (TB) or design storm duration (D). However, changes in the 4 variables of DPSBAR, PROPWET, DPLBAR and URBEXT will cause a change in TB and thus D, whilst a change in SAAR will cause a change in D. Care must be taken when D or TB are changed, as the convoluted hydrograph changes its Row numbers in the "Design" worksheet. The following procedure may be adopted.

1. Change, for example, DPSBAR in the "Design" worksheet to 10 m/km.
2. Activate the "Run" worksheet and the return period Spin button. The "Runoff" curve [Series Collection(2)] in the "Computation" chart will now appear corrupted, since it plots the row values of the previous hydrograph.
3. Activate this curve with a right click and use the Excel commands [Source Data, Series, Runoff].
4. Use the Cursor to highlight the [X values] and the worksheet will change to the "Design" sheet cells of the previous hydrograph. One may now enter the new row of hydrograph time values by highlighting them in the "Design" worksheet.
5. The [Y values] represent the hydrograph depth values, and these may be edited as above.

6. Activating [OK] for this [Source Data] window will save and plot the new hydrograph.

Note that the Scroll bar button should not be activated until the "Runoff" curve has been edited.

The present code has been arbitrarily set to run a maximum duration for the design hydrograph of 100 hours. This can of course be extended as the worksheet can easily accommodate more data. However, it is realised that for extended storms, the time intervals for the UH and design storm profile are better set to longer values than 1 hour. The existing code must be edited for variable time intervals.

6.1 RAINFALL-RUNOFF USING THE SCS METHOD

6.2.1 Summary

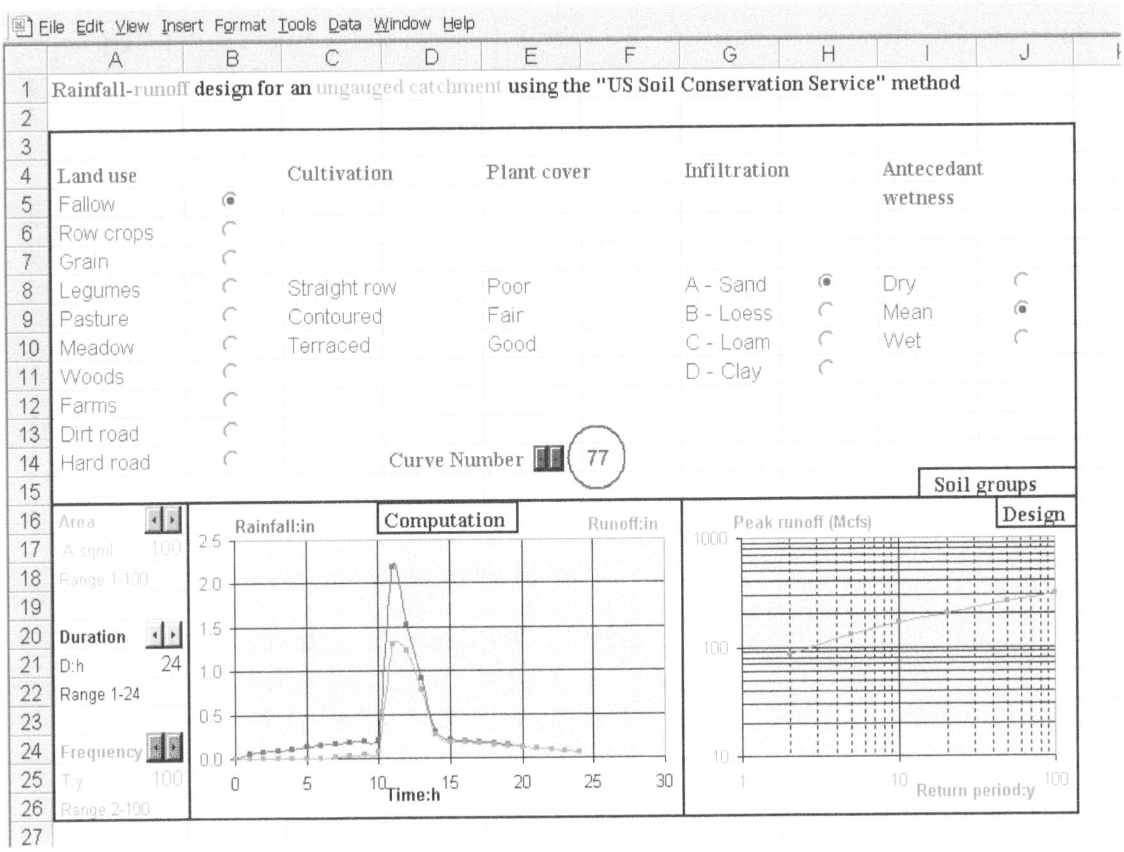

Figure 6.2.1: The "Run" worksheet for the "SCS.xls" project

In contrast to the FEH rainfall–runoff method for a gauged catchment in a small country, we now consider the US Soil Conservation Service (SCS) method for an ungauged catchment in a large continent. The presentation follows that given by Roberson et al (1997), and is limited to the direct conversion of rainfall to runoff without the use of a gauged, unit hydrograph. The FEH method may be contrasted with the SCS method used herein in terms of the design storm profile and the rainfall loss methods of estimation.

The SCS design storm profile differs principally with FEH in that the frequency used is the same as the design runoff flood frequency, and the storm may have any duration. The latter is due to the many climatic zones that may exist in the US. For example, an extreme rainfall generated during several hours of a moving tropical storm may take several days to occur for a slow frontal passage. The rainfall loss method is also more detailed because of the much larger area; it has been modelled herein as a "knowledge based expert system" (Krishnamoorthy, 1996). The design, flood peak runoff is finally determined against return period (Figure 6.2.1). Note that since the method was originally defined in the US system of units, the latter are maintained (1 inch = 25.4mm, 1mile = 1.609344km).

6.2.2 Code operation

Soil groups

Land use	Cultivation	Cover	Sand soil (A)	Loess soil (B)	Loam soil (C)	Clay soil (D)
Fallow			77	86	91	94
Row crops	Straight row	Poor	72	81	88	91
	Straight row	Good	67	78	85	89
	Contoured	Poor	70	79	84	88
	Contoured	Good	65	75	82	86
	Terraced	Poor	66	74	80	82
	Terraced	Good	62	71	78	81
Woods		Poor	45	66	77	83
		Fair	36	60	73	79
		Good	25	55	70	77
Roads	Hard surface		74	84	90	92

Table 6.2.1: Excerpt from the SCS curve number knowledge base

The soil-cover complex method of the SCS (1964) defines hydrological soil groups in terms of a "curve number". (The latter is approximately analogous to the "Standard percentage runoff" for the 29 HOST soil classes used in the FEH method.) Table 6.2.1 illustrates how a particular land use may be classified into 3 cultivation or treatment types, 3 cover types, and 4 hydrologic soil group types.

The land use type has a total of 10 alternatives. The cover type is defined as:

> Poor :- heavily grazed with < 0.5 of area with plant cover.
> Fair :- moderately grazed with plant cover between 0.5 and 0.75 of area.
> Good :- lightly grazed with > 0.75 of area with plant cover.

The hydrological soil group types are based on their rate of infiltration, thus:

Sand soil: 0.30 to 0.45 in/hr - deep, well-drained sands or gravel.
Loess soil: 0.15 to 0.30 in/hr - moderately deep soils such as loess and sandy loam.
Loam soil : 0.05 to 0.15 in/hr - moderately fine grained such as clay and sandy loams.
Clay soil: 0.00 to 0.05 in/hr - fine grained clay soils.

The SCS "knowledge base" (a term used in Artificial Intelligence for defining the information for an "Expert system") which is excerpted in Table 6.2.1 has a total of 128 curve numbers. (This may be compared with the 29 HOST classes used in FEH, or say 30 classes if the area of urban coverage, URBEXT, is included.)

The curve numbers (CN) may also be influenced by an "Antecedant Moisture Condition" (similar to the "Catchment Wetness Index", CWI, used in FEH). The values given in

Table 6.2.1 are for a mean condition that most frequently occurs before large storms. A polynomial can be used to transform the mean condition (curve number, x) to a dry or wet condition (y) as follows:

Dry:	$y = 0.0000484x^3+0.000185x^2+0.4729x$	$R^2 = 0.999$
Wet:	$y =-0.00000352x^4+0.000752x^3-0.0592x^2+2.907x$	$R^2 = 0.998$

(Note that this transformation is analogous to the variable of "Dynamic percentage runoff" with CWI, as used in FEH.)

The curve number expert system

There are thus a total of 384 (128 * 3) curve numbers. Figure 6.2.1 illustrates how each of the 10 "land use" types may be selected in sequence, and categorised according to the maximum of 3 cultivation types, 3 plant cover types, 4 infiltration types and 3 antecedant moisture types. The categories are simply selected by activating the adjacent Option buttons, and the relevant CN will display in the circle. An arbitrary CN may also be selected without reference to the knowledge base using the Spin button adjacent to the circle.

The CN selection method used herein is similar to that used for an "expert system" in the study of Artificial Intelligence. A series of 5 category questions are answered with Option buttons, and the CN is displayed. The method is however simplified at the first level (the land use level), in that the code gives only the relevant categories to be selected. For example, the "Fallow" land use requires selections from only the infiltration and antecedant moisture categories.

Computation and Design areas

In addition to changing the CN, the "CN" Spin button control runs code in the Sheet1 (Run) object to compute the rainfall and runoff hyetographs. The latter are also recomputed when changes are made to the catchment area (A), storm duration (D) and frequency (T). These changes are made using the associated Spin buttons in the control panel area of the "Run" worksheet.

The Design area is changed using the "Frequency" Spin button. Whenever a change has been made in any of the other controls (including the CN itself), the "Frequency" Spin button must be run throughout its range to recompute the design curve. To speed this process, the "Frequency" Spin button code has been designed so that no worksheet changes are required when data is read or written. The Spin button therefore changes continuously and rapidly (i.e. it spins) when activated.

6.2.3 Procedure

Design storm depth

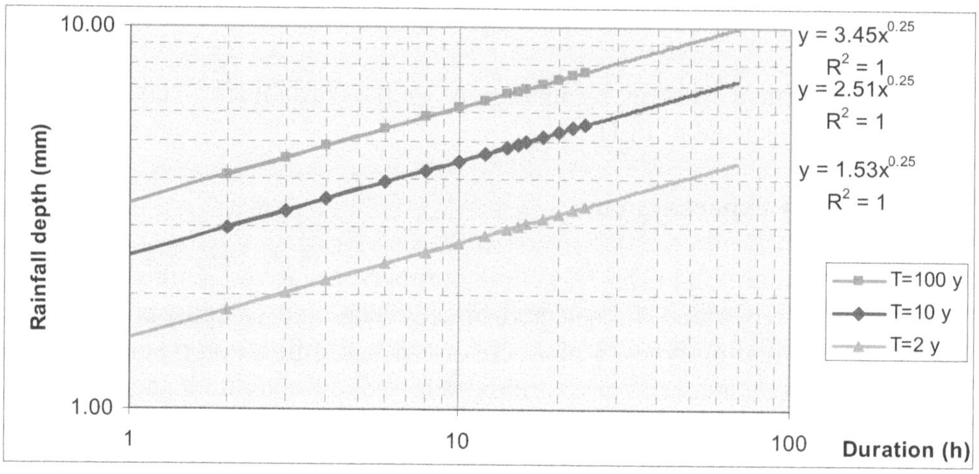

Figure 6.2.2: A depth, duration and frequency curve (DDF curve)

The US Weather Bureau have published a "Rainfall Frequency Atlas for the US" (1961). Nationwide and State maps are drawn showing lines of equal rainfall depth (isohyets) for different return periods and storm durations. DDF curves may be synthesised from the maps, and an example for a location in eastern Arkansas (Roberson et al, 1997) is given in Figure 6.2.2. The data are plotted on a logarithmic scale, and a growth coefficient for use with any frequency is calculated as:

$$\text{growth coefficient} = 0.4851\text{Ln}(T) + 1.2676 \qquad R^2 = 0.987$$

whence: $$\text{rainfall depth} = \text{growth coefficient} * D^{0.25}$$

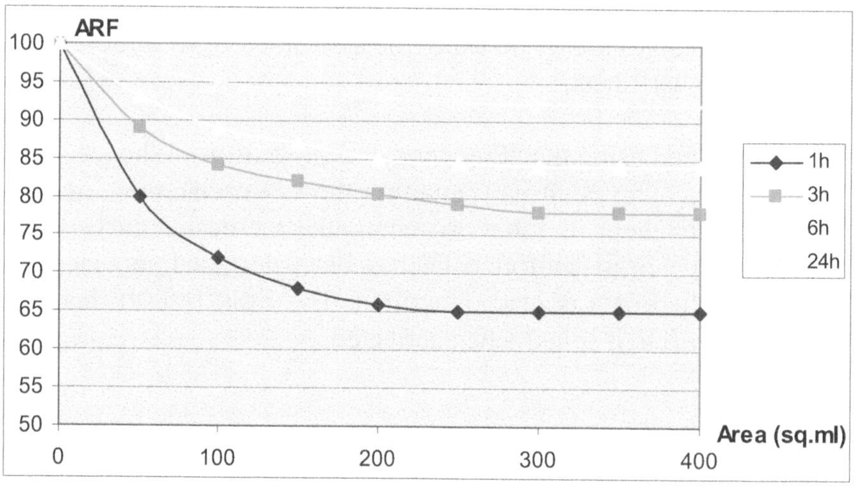

Figure 6.2.3: Areal reduction factor for point rainfalls

Figure 6.2.3 illustrates the US Weather Bureau design curves for adjusting point rainfall to average depths over an area. These curves may be simulated analytically as a quadratic trend surface, thus:

$$ARF = 85.79 + 3.677D - 0.1439T - 0.1245D^2 + 0.00022T^2 + 0.0017DT$$

The "goodness of fit" indicator for this trend is given by $R^2 = 0.911$.

Design storm profile

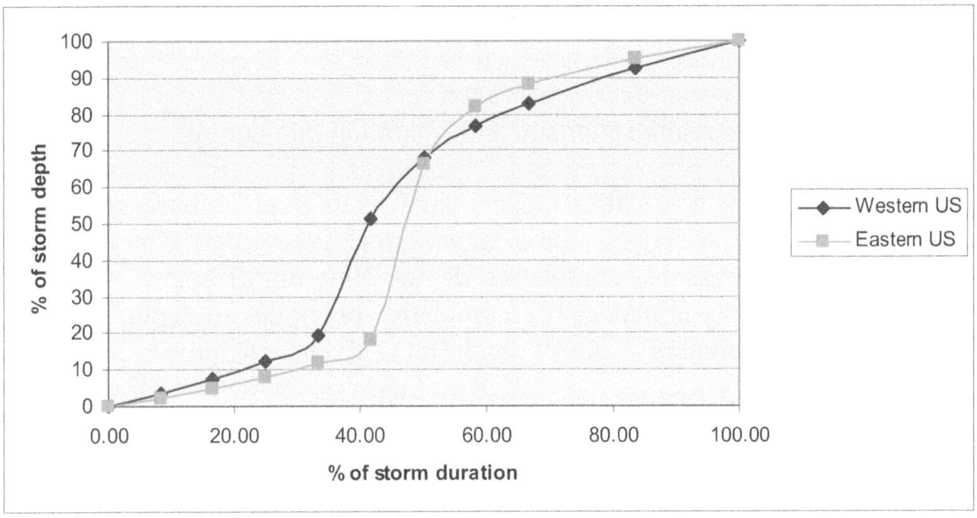

Figure 6.2.4: SCS profiles for a 24 hour duration storm

Although many design storm profiles exist for the US, the SCS profile for the Eastern US (Figure 6.2.4) is used for this example. (Note that this profile differs from the FEH representation, as it represents the cumulative depth from the start to the finish of the storm.) The profile may be analytically represented by 3 individual quadratic polynomials, thus:

$\%D \leq 41.67$:	$\%P = 0.0061\%D^2 + 0.1719\%D$	$R^2 = 0.998$
$41.67 < \%D < 58.33$:	$\%P = -0.234\%D^2 + 27.234\%D - 710.4$	$R^2 = 1$
$D > 58.33$:	$\%P = -0.0064\%D^2 + 1.4346\%D + 20.275$	$R^2 = 0.998$

Approximately 60% of the rainfall occurs in the central 16% of the duration. The hyetograph in the "Run" worksheet illustrates the asymmetry and high kurtosis of this design profile.

Runoff hyetograph

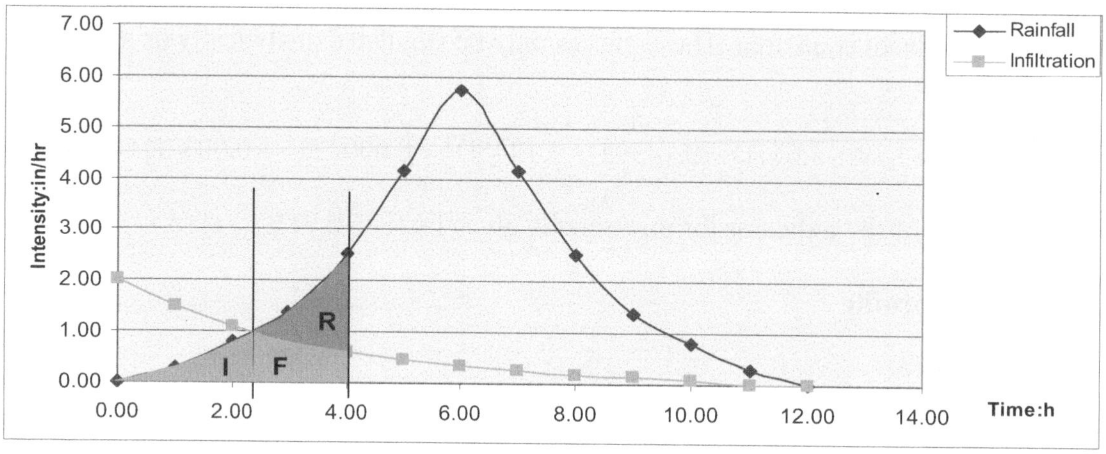

Figure 6.2.5: Hypothetical rainfall intensity and infiltration rate curves

The underlying reasoning for estimating a runoff hyetograph is based on the rainfall intensity and infiltration rate curves for a catchment (Figure 6.2.5). The area subtended by a rate curve will represent a cumulative depth. Thus runoff begins when an initial abstraction (I) has been accumulated. If a cumulative precipitation depth, P, occurs after runoff begins, then the cumulative runoff depth (R) is given by (Figure 6.2.5):

$$R = P - F$$

where F is the cumulative infiltration depth. The SCS method assumes that the infiltration rate reaches zero before the end of a storm, thus $R/P \rightarrow 1$. Similarly, if the total infiltration is given by F_{max}, then $F/F_{max} \rightarrow 1$ before the end of the storm. The assumption is therefore made that:

$$F/F_{max} = R/P$$

However, since there is an initial abstraction , I, the functional similarity becomes:

$$F/(F_{max} + I) = R/(P - I)$$

Data analyses for small catchment areas has shown that:

$$I = F_{max} / 4 \qquad \text{or} \qquad I = 0.2(F_{max} + I) = 0.2S$$

Substituting for I in the functional similarity equation then gives:

$$R = (P - 0.2S)^2 / (P + 0.8S)$$

The curve number is finally defined as a percentage, so that R = P at S = 0, thus:

$$CN = 1000/(10 + S)$$

When CN = 100, S = 0 and R → P, and when CN = 0, S → ∞. It is thus seen that the CN is an inverse function of the sum of the initial abstraction and infiltration. It is analogous to the FEH percentage runoff for a HOST soil condition.

Discussion

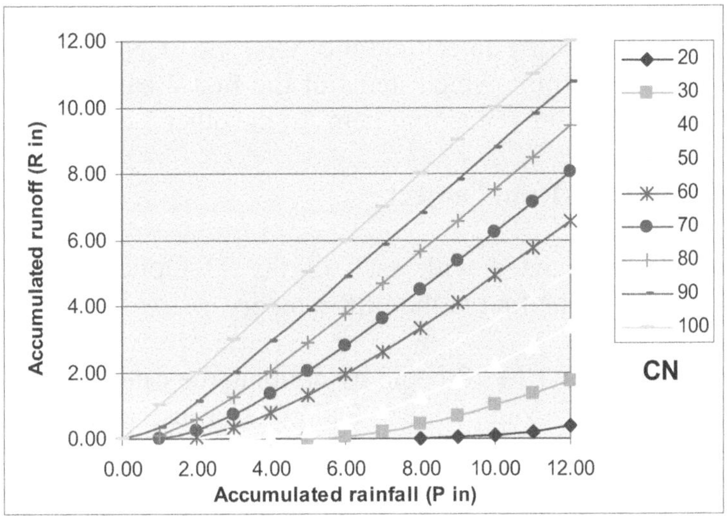

Figure 6.2.6: Accumulated runoff depths for various curve numbers

The curve number influence on runoff is synthesised in Figure 6.2.6. It is apparent how the runoff decreases and is delayed with decreasing CN. However, since the theory assumes that the infiltration rate becomes zero during a storm, it becomes inaccurate for storms of several days duration (Roberson et al, 1997). Furthermore, since the I = 0.2S relation was obtained for small catchment areas only, the method should only be used for small areas up to about 10 square miles. Nevertheless, the method has been applied to larger areas, and for this reason the Excel project herein uses a maximum area of 100 square miles with a maximum storm duration of 24 hours.

The FEH rainfall losses method uses a standard percentage runoff equivalent to the SCS CN runoff method, and a dynamic percentage run off for CWI equivalent to the SCS antecedant moisture condition. FEH also uses a dynamic percentage runoff based on the storm magnitude. This condition may be considered equivalent to the initial abstraction losses of the SCS method, which is based on the storm's cumulative retention of precipation ($I = F_{max} / 4$). The loss methods are therefore comparable.

6.2.4 Modules

The Soil group expert system consists of 23 items in 5 categories. The latter are land use (10 items), cultivation (3 items), cover (3 items), infiltration (4 items) and wetness (3items). The Option buttons for each category are first grouped using the Excel command [Properties, Groupname]. Associated code is contained in the Private subroutines of the Sheet1 (Run) object and the "CN1" to "CN10" modules. In general, the Private subroutines determine which items of the first 2 categories are selected and the "CN" modules determine the remaining items of the other 3 categories.

Subroutines in the "Sheet1 (Run)" object

Private subroutines are associated with each of the 23 Option button events. They increase in complexity from the first to the fifth category.

The code for Option buttons 1 to 10 begin by defining the categories and items that are relevant for the land use. For example, the "Fallow" land use category does not require items from the second or third category. The code is thus:

```
Private Sub OptionButton1_Click()
OptionButton11.Visible = False: OptionButton12.Visible = False:
OptionButton13.Visible = False: OptionButton14.Visible = False:
OptionButton15.Visible = False: OptionButton16.Visible = False
```

The code can determine which item of category 2 is selected, and runs the appropriate code in the CN subroutines. For example, the "Grain" land use category uses the following code:

```
Private Sub OptionButton3_Click()
OptionButton11.Visible = True: OptionButton12.Visible = True:
OptionButton13.Visible = True: OptionButton14.Visible = True:
OptionButton15.Visible = False: OptionButton16.Visible = True
If OptionButton11.Value = True Then CN3_1_1: CN3_1_2
If OptionButton12.Value = True Then CN3_2_1: CN3_2_2
If OptionButton13.Value = True Then CN3_3_1: CN3_3_2
End Sub
```

The code for Option buttons 11 to 13 need only determine which Land use buttons are selected and run the appropriate CN subroutine. For example:

```
Private Sub OptionButton11_Click()
If OptionButton2.Value = True Then CN2_1_1: CN2_1_2
If OptionButton3.Value = True Then CN3_1_1: CN3_1_2
If OptionButton4.Value = True Then CN4_1_1: CN4_1_2
If OptionButton5.Value = True Then CN5_1_1: CN5_1_2: CN5_1_3
End Sub
```

The codes for Option buttons 14 to 23 must however determine which buttons are selected for both the land use and cultivation categories. The code therefore uses a double If condition. For example:

```
Private Sub OptionButton14_Click()
If OptionButton2.Value = True Then
If OptionButton11.Value = True Then CN2_1_1: CN2_1_2
End If
```

These multiple If conditions occur frequently in this project, and should be interpreted in English as:

If a value is true <u>and</u> if another value is true then run the relevant subroutine.

"CN"

Each curve number module refers to one of the 10 land use type items. The separate subroutines in each module refer to a "mean infiltration" value for the curve numbers, as given in Table 6.2.1. There are therefore a total of 32 subroutines, each with a unique curve number for the mean infiltration item (A, B, C, D). The "CN2" module is now described, since it is a module with the most categories.

For "CN2", there are 3 cultivation items and 2 cover items making a total of 6 sets of mean infiltration curve numbers. There are therefore 6 subroutines that are named according to the cultivation or cover category. For example, subroutine "CN2_1_1" refers to the second land use item (Row crops), the first cultivation item (straight row, Option button 11) and the first cover item (Poor, Option button 14). And subroutine "CN2_3_2" refers to the second land use item (Row crops), the third cultivation item (terraced, Option button 13) and the second cover item (Good, Option button 16). Note that the cultivation selection has been made in the Private subroutines.

Since each "CN2" subroutine has a predetermined cultivation item, the code begins with a single If statement to determine the cover item. The End If part of this statement is at the end of the subroutine. This first If condition is included in a multiple If statement to determine the wetness and infiltration items. The curve number is then evaluated using the wet or dry polynomials or simply the mean condition declared at the start of the subroutine.

"Hydrology"

This module contains 3 subroutines that are similar to those used in the FEH project. The first subroutine, "StormProfile", is in 3 steps, namely:

1. Define the time series, which is given in hourly increments using the Excel [Edit, Fill, Series] commands.

2. Calculate the hourly cumulative profile using the 3 polynomials derived for the profile.
3. Calculate the hourly hyetograph by subtracting sequential cumulative values.

The next subroutine, "Runoff" is again in 3 steps:

1. Calculate the hourly cumulative runoff using the SCS formula with cumulative precipitation.
2. Delete the spurious values from formulae. Note that the method is semi-empirical and the runoff formula gives spurious values at the start of a storm. These values decrease in magnitude to zero, until a minimum value indicates the start of runoff.
3. Calculate the hourly runoff hydrograph by subtracting sequential cumulative values, and multiplying the resulting hyetograph with the area.

Finally, the "EVCurve" subroutine selects the peak runoff value for each of the 2, 5, 10, 50 and 100 year return periods.

Note that all of the VB Range objects in this module are preceeded by the associated Sheets object. The Sheets objects were inserted after the code was written using the Excel [Edit, Replace, Replace All] commands. It was done to prevent the code from selecting worksheets during runtime, and thus enabling the Spin buttons to operate continuously. It is of course easier to first develop code by changing worksheets.

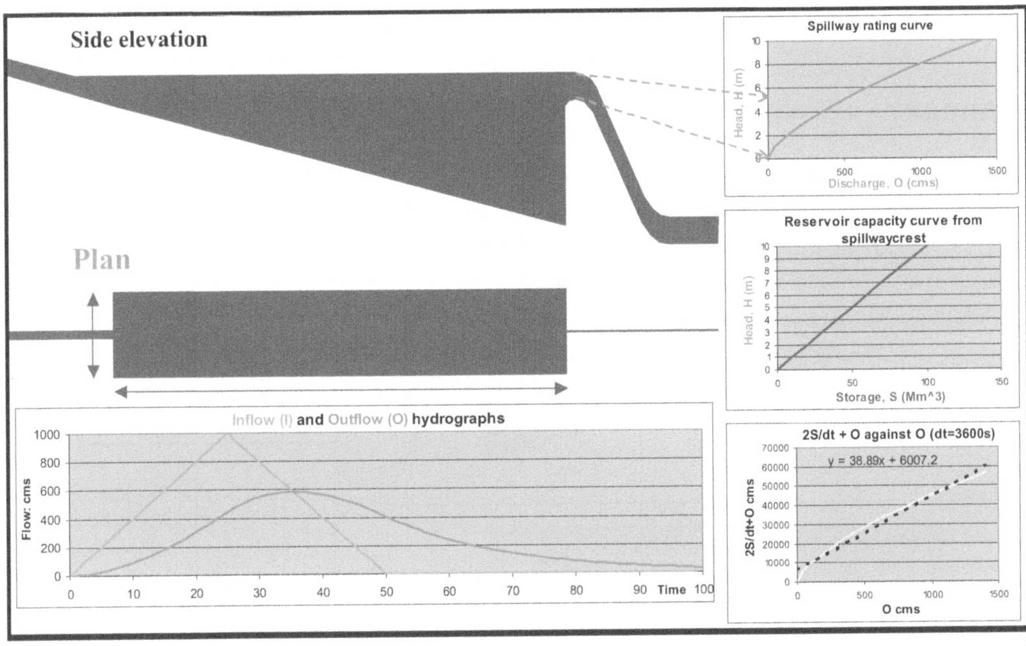

Reservoir design

7.1 RESERVOIR ROUTING OF FLOODS

7.1.1 Summary

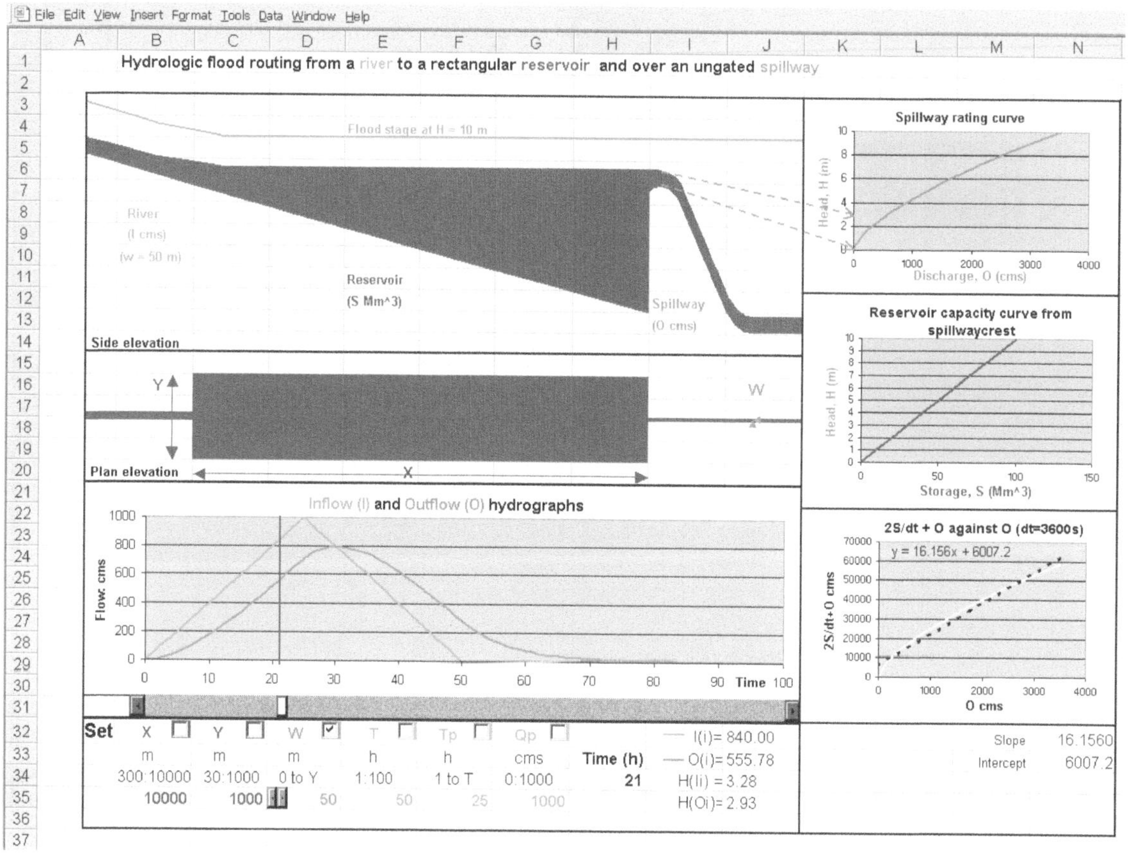

Figure 7.1.1: The "Run" worksheet for the "ResRoute.xls" project

This project evaluates the inflow and outflow hydrographs for a flood that passes through a reservoir (Figure 7.1.1). It illustrates how the inflow is attenuated, due to the temporary storage of water in the reservoir. This temporary storage is released after the flood peak has passed. The resulting outflow hydrograph for the flood is therefore less dangerous than the inflow hydrograph.

For simplicity, the reservoir is designed to be rectangular in plan (throughout its depth) and triangular (to the spillway crest level) in side elevation. The inflow is from a river of constant width (w = 50 m), and the outflow is over an ungated spillway of variable width (W). A general hydraulic solution for this example problem would include unsteady accelerating and decelerating flow in the reservoir. Since the reservoir area and depth are large, the unsteady flow is small and may be neglected. The reservoir surface is therefore assumed horizontal, and only the flow continuity equation is used to evaluate a solution. The solution method is known as "hydrologic routing", and is sometimes called "level pool routing".

7.1.2 Code operation

The control panel for this project is at the bottom of the "Run" worksheet. There are 2 main sets of controls. These are the "Run" Scroll bar which selects the hydrograph time, and the 6 "Set" Check boxes which enable the variation of the reservoir length and breadth (X and Y), the spillway width (W), the inflow hydrograph time base (T), time to peak (Tp) and peak flow (Qp).

When a "Set" Check box is clicked on, a Spin button becomes visible that enables the associated variable to be altered. Additionally, the Check box control code in the Sheet1 (Run) object resets the "Run" Scroll bar to zero, since a change in project variables will require a new hydrograph.

The project variables are set to an arbitrary but practical maximum in their respective Spin button properties object. The reservoir length (X) has a range from 300 m to 10 km. The lower limit is simply a computational limit below which the code fails. The reservoir breadth (Y) has a range from 30 m to 1 km, and the lower limit is again determined by code failure. The spillway width (W) will not exceed the reservoir width, so this is coded in the the SpinButton3 control subroutine as:

```
SpinButton3.Max = Sheet1.Cells(35, 3).Value
```

The spillway width may indeed be zero, in which case the reservoir is being used totally for detention. However, if the flood level exceeds that permitted (i.e. when the flood stage = 10 m), a Message Box warning is evoked in the "DrawFlood" subroutine of the "Drawing" module.

The input hydrograph timebase (T) has a range from 1 hour to 100 hours, and should be greater than Tp. A Message box warning is invoked in the "InputFlow" subroutine of the "Flows" module if T > Tp. Finally, the input peak flow variable (Qp) has a range from 10 m^3/s to 1000 m^3/s in 10 m^3/s increments.

There are 2 main conditions of code operation. The first is when the spillway width is zero, and the input flow is totally detained in the reservoir. For this case, the following subroutines are run by all the Spin button control codes:

"EraseSpillway; StorageCurve; Inputflow; Outputflow; Flood"

The second condition occurs when a spillway flow occurs, and the following subroutines are then run by all the Spin button control codes:

"DrawSpillway; StorageCurve; MacroTrendline; Inputflow; Outputflow; Stage"

The above subroutines compute all the necessary statistics, and the "Run" Scroll bar control simply plots the water stages, from these statistics, at a particular time.

7.1.3 Theory

River inflow rating curve

It is necesssary to compute the river stage (H(i)) from the known inflow discharge (I) when drawing the river level for any particular hourly time (i). In river hydraulics, the dimensionless relationships for steady open channel flows are adapted from pipe flow experiments (Roberson et al, 1997). For example, the head loss for turbulent flow in pipes is given by the Darcy – Weisbach formula as:

$$\text{Head loss} = f * L/D * V^2/2g$$

where f is a pipe resistance coefficient (which may be interpolated from a Moody diagram), L is the pipe length, D the pipe diameter, V the velocity of flow in the pipe and g the gravitational acceleration (9.81 m/s^2). This formula is adapted to channels by noting that the head loss in an open channel is given by (S * L) where S is the channel slope and L the channel length. The pipe diameter (D) is replace by 4R for a channel, where R is called the "hydraulic" radius denoted by the area/perimeter ratio. By equating the head losses, the channel velocity is solved as:

$$V = (8gRS/f)^{0.5}$$

If a wide channel is assumed, then $R \approx H$ and:

$$I = V * \text{area} = (8gS/f)^{0.5} * w * H^{1.5}$$

where w is the channel width. By choosing physical values of $S = 0.001$, $f = 0.01$ then:

$$H = 0.5*(I/w)^{0.667}$$

Thus H(i) can be evaluated for any discharge in this example. Since a maximum, hypothetical flood stage of 10 m is taken for this project and w = 50 m, the wide channel approximation is used throughout.

Spillway outflow rating curve

The discharge (O) over an ungated spillway is controlled mainly by the head (H) above the spillway crest, thus:

$$O = C(2g)^{0.5} * WH^{1.5}$$

where C is a dimensionless coefficient which varies with the reservoir depth and the spillway shape at the crest (Roberson et al, 1997). For deep reservoirs, $C \approx 0.5$, thus:

$$O = 2.215 * WH^{1.5}$$

and

$$H = [O/(2.215*W)]^{0.667}$$

Reservoir routing

When applying flow continuity to a reservoir, the rate of inflow (I) minus the rate of outflow (O) is equal to the rate of change of the volume of water (S) stored in the reservoir. Thus:

$$I - O = dS/dt$$

Expressing this in finite difference form, then:

$$(I_i + I_{i+1})/2 - (O_i + O_{i+1})/2 = (S_{i+1} - S_i) / \Delta t$$

where i is an hourly time. This equation may be rewritten as:

$$I_i + I_{i+1} + 2S_i/\Delta t - O_i = 2S_{i+1}/\Delta t + O_{i+1}$$

It is solved for each O_i, noting that the inflow is defined by the input hydrograph and thus all the hourly I_i values are known. The solution is as follows:

1. At the first hourly time of zero (i = 1), $I_1 = 0$, $S_1 = 0$ and $O_1 = 0$. Therefore $(2S_2/\Delta t + O_2) = I_2$ and I_2 is known.
2. From a "storage curve" of $(2S/\Delta t + O)$ versus O (described below), O_2 can be evaluated since $(2S_2/\Delta t + O_2)$ is known.
3. At the second hourly time (i = 2), $(2S_2/\Delta t - O_2)$ is evaluated by subtracting $2O_2$ from $(2S_2/\Delta t + O_2)$.
4. The left hand side of the equation may then be evaluated to give the right hand side, $(2S_3/\Delta t + O_3)$.
5. From a curve of $(2S/\Delta t + O)$ versus O, O_3 can be evaluated.
6. This recursive procedure is repeated for all i.

Storage curve

This curve is derived from the spillway-rating curve (H versus O, as shown in the "Run" worksheet and given above) and a computed reservoir capacity curve (H versus S, as shown in the "Run" worksheet). The latter is simply derived by computing the cubic volume for a given temporary storage head. Since both curves are a function of the same head (H) above the spillway crest, then a curve of $(2S/\Delta t + O)$ versus O can be derived. The Δt increment is taken as 1 hour, thus 2S is divided by 3600 seconds to attain the same units as the outflow (m^3/s).

The above computations are made by the "StorageCurve" subroutine in the "Flows" module, and written to the "Storage" worksheet. The interpolation of this curve to find O_{i+1} (given $2S_{i+1}/\Delta t + O_{i+1}$) requires a trendline for the curve. A linear regression trendline was used throughout with small error. And the regression coefficients for this trendline were abstracted from the chart using the "MacroTrendline" subroutine in the "Flows" module.

7.1.4 Modules

Two modules are used for this project. The "Flows" module computes the hydrograph statistics and writes them to the "Storage" and "Design" worksheet, and the "Drawings" module draws the moving objects on the "Run" worksheet. The associated subroutines are now described.

"Flows"

The first of 6 subroutines in this module, "StorageCurve", writes the data to the "Storage" worksheet. The worksheet columns are written as follows:

1. Column A contains the H values from 0 to 10.
2. Column B contains the associated outflow values that are calculated from the spillway rating curve.
3. Column C contains the associated reservoir storage (H * X * Y) in Mm^3.
4. Column D contains the $(2S/\Delta t + O)$ calculation.

These columns enable a plot of the Storage curve chart in the "Run" worksheet, and the approximated trendline. The second subroutine "MacroTrendline" parses the trendline equation into its slope and intercept coefficients, for use in interpolation. The parsing is done with the aid of the Excel [TextToColumns] command.

The third subroutine,"InputFlow", computes the input hydrograph and writes it to the "Design" worksheet. The first 4 worksheet columns are written as follows:

1. Column A contains the time values from 0 to 100 in hours.
2. Column B contains the "i" number of the hourly value from 1 to 101.
3. Column C contains the input hydrograph from 0 hours to T hours.
4. Column D contains the input hydrograph summation for part of the finite difference equation, namely $I_i + I_{i+1}$.

The fourth subroutine, "Outputflow", computes the parameters necessary for evaluating the outflow and writes them to the "Design" worksheet. The subroutine follows the recursive computation detailed above, and the worksheet columns appear as follows:

1. Column E contains the $(2S_i/\Delta t - O_i)$ term.
2. Column F contains the $(2S_{i+1}/\Delta t + O_{i+1})$ term.
3. Column G contains the O_i term.

The fifth subroutine, "Stage", computes the water levels for the river input (HI(i)) and spillway output (HO(i)) using the rating curves defined above. And the sixth subroutine computes the reservoir stage (H(S)) when there is no spillway. The H(S) stage is simply calculated by dividing the cumulative storage from the inflow by the reservoir area. These level values are then used in the "Drawings" module.

"Drawings"

There are 6 subroutines in this module, and they all convert the physical metric units to worksheet drawing units in order to draw the relevant objects. The first 2 subroutines are "ChangeLength" and ChangeBreadth". They each begin by converting the length (X) and breadth (Y) variables to worksheet units. The reservoir plan elevation is then altered according to the desired setting of X and Y.

The "DrawStage" and "DrawFlood" subroutines enable the drawing of water levels in the side elevation for the normal and closed spillway (W = 0) operation respectively. They each begin with the computed result of HI(i) and HO(i) or HI(i) and HS(i) from the "Design" worksheet. These values are written to the control panel of the "Run" worksheet to assist in the interpretation of the moving objects. The H values are then converted to worksheet units to enable the drawing of the river, reservoir and spillway (for "DrawStage") levels.

The fifth subroutine, "EraseSpillway", is used for the closed spillway condition. It simply changes the freeforms of all spillway objects to the `visible = false` condition. The spillway text and dimension arrows are erased by changing their colours to white.
The sixth subroutine, "DrawSpillway", reverses the commands of "EraseSpillway". In addition, this subroutine draws the spillway dimensions in the plan elevation.

Numerical check

The following values for the 6 variables are used for this check:

X = 10000 m; Y = 1000 m; W = 20 m; T = 50 h; Tp = 25 h, Qp = 1000 m^3/s

The storage values written to the "Storage" worksheet can be checked for, say, the H = 10 m level in 3 steps, thus:

1. $O = 2.215*WH^{1.5} = 2.2125*20*10^{1.5} = 1400.9\ m^3/s$
2. $S = (H*X*Y)/1000000 = 10*10000*1000/1000000 = 100\ Mm^3$
3. $2S/3600 + O = 200*1000000/3600 + 1400.9 = 56{,}956\ m^3/s$

And the trendline regression may be roughly checked against the value in step 3:

$$2S/3600 + O = 38.89*1400.9 + 6007.2 = 60488\ m^3/s$$

This value is a few percent higher than the step 3 value. A more accurate polynomial trendline may be used to better model the storage curve, and is left as a student exercise.

For the inflow hydrograph values written to the "Design" worksheet, the hourly increments are simply checked as

$$Qp/Tp = 1000/25 = 40\ m^3/s$$

The recursion of the finite difference equation is now checked according to the same steps given in the theory. Note that the trendline approximation is used in place of the actual storage curve.

1. $I_1 = 0$; $I_1 + I_2 = 40$; $O_1 = 0$; $2S_1/\Delta t - O_1 = 6007.2$; $2S_2/\Delta t + O_2 = 6047.2\ m^3/s$
2. $O_2 = (6047.2 - 6007.2)/38.89 = 1.0285\ m^3/s$
3. $2S_2/\Delta t - O_2 = 2S_2/\Delta t + O_2 - 2 * O_2 = 6047.2 - 2.057 = 6045.14\ m^3/s$
4. $2S_3/\Delta t + O_3 = 120 + 6045.14 = 6165.14\ m^3/s$
5. $O_3 = (6165.14 - 6007.2)/38.89 = 4.0613\ m^3/s$

We can finally check the rating curves in the "Design" worksheet according to the following steps:

1. $HO_2 = [O/(2.215*W)]^{0.667} = [1.03/(2.215*20)]^{0.667} = 0.08$ m
2. $HI_2 = 0.5*(I/w)^{0.667} = 0.5*(40/50)^{0.667} = 0.43$ m
3. $S_2 = (I*3600)/1000000 = 0.14\ Mm^3$
4. $S(cumulative) = 0.14\ Mm^3$
5. HS = (S(cumulative) * 1000000) / (X * Y) = 0.14 * 1000000/10000000 = 0.014 m

These calculated values agree with the computed values. However, the tedium involved in checking just a few hourly increments emphasises our dependence on computers.

7.2 RESERVOIR STORAGE DESIGN

7.2.1 Summary

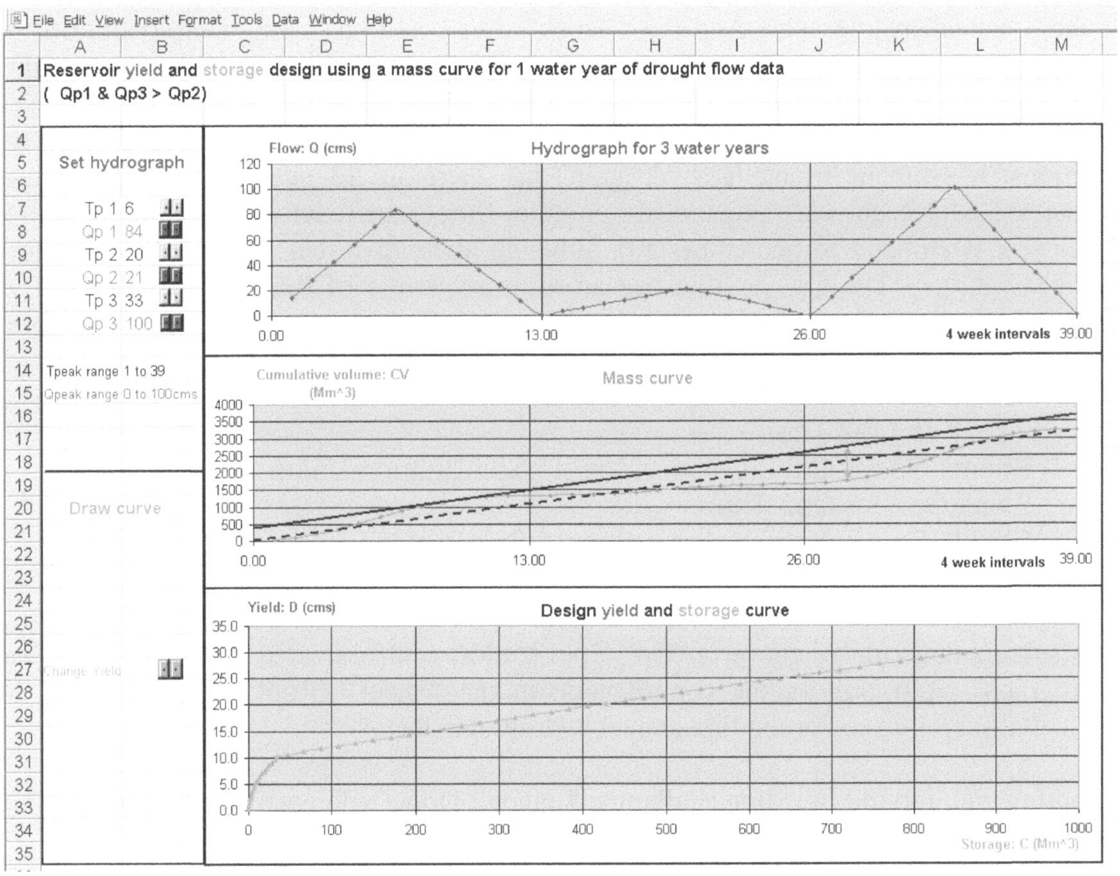

Figure 7.2.1: The "Run" worksheet for the "ResStore.xls" project

This project provides an estimate for the storage requirements of a reservoir that is to be designed for a natural stream. Initially, the total streamflow records are scanned to find the most severe drought condition. The latter is simulated herein by use of 3 synthetic triangular hydrographs for a 3-year flow record, in which the second year is the drought condition. The cumulative volume of flow is then plotted against time (the "mass" curve), and the slope of the mass curve at any point is the stream discharge at that time.

A straight line (shown dashed) drawn from the beginning to the end of the 3-year mass curve would represent an available, steady discharge for the period (Figure 7.2.1). A second straight line (shown solid) is drawn parallel to the steady discharge and tangential to the first year of the mass curve. Immediately after this point of tangency, the discharge decreases and storage is needed to maintain the discharge. The required storage for the entire drought is given by the maximum difference between the tangential line and the mass curve. (Note that it is assumed that the reservoir has filled antecedently at the time of tangency.) The design curve for any steady discharge (the "yield") is computed by lowering the slope of the tangential line to zero, and computing each associated storage.

7.2.2 Code operation

There are 2 areas of operation in the control panel at the left of the "Run" worksheet. The "Set hydrograph" area is used to adjust the peak flow (Qp) and time to peak (Tp) for each of the 3 annual hydrographs. And the " Draw curve" area has a single Spin button to generate the design "Yield-Storage" curve. These areas are now described in more detail.

Each year is defined as a multiple of 4-week intervals, to avoid the inequality of monthly durations. A synthetic triangular hydrograph has therefore 13 intervals, and Tp varies from 1 to 12 for each year. The annual minimum flow is artificially set at the end of the water year (October 1 to September 30), although alternative real data may be used for the mass curve. The Qp variable has an arbitrary range of 0 cms to 100 cms. By definition, Qp1> Qp2 and Qp3 > Qp2. A "Boundary Condition" subroutine is therefore invoked in the "Hydrograph" module to ensure this condition, thus:

```
If Qp1 <= Qp2 Then
Sheet1.SpinButton2.Value = Sheet1.SpinButton4.Value + 1
ElseIf Qp3 <= Qp2 Then
Sheet1.SpinButton6.Value = Sheet1.SpinButton4.Value + 1
End If
```

This subroutine is run whenever a "Qp" Spin button control is activated. For each "Set hydrograph" Spin button, the control code in the "Sheet1 (Run)" object invokes 6 sequential subroutines. For the first annual hydrograph, they are:

"InputInterval; InputFlow1; InputColumns; Tangent; DrawDifference; YieldCurveMax"

In summary, the subroutines calculate the relevant annual hydrograph shape, the mass curve, the tangential slope, the storage requirement, and the maximum yield and storage coordinates on the final design curve. All of these parameters are plotted on the 3 charts of the "Run" worksheet.

The "Draw curve" Spin Button invokes fewer subroutines, since the hydrograph is now maintained constant. The "Yield" ordinate is varied by decrementing the highest value of the cumulative volume by 50 Mm^3. Note that at the lower yields, the "storage" arrow on the mass curve moves nearer to the tangential point until it reaches the tangential point at zero discharge. The subroutines invoked are:

"TangentVar; DrawDifference; YieldCurveDown"

In summary, the subroutines calculate the new tangential slope, the new storage requirement, and the final design curve. These parameters are plotted on the 2 lower charts of the "Run" worksheet.

7.2.3 Discussion

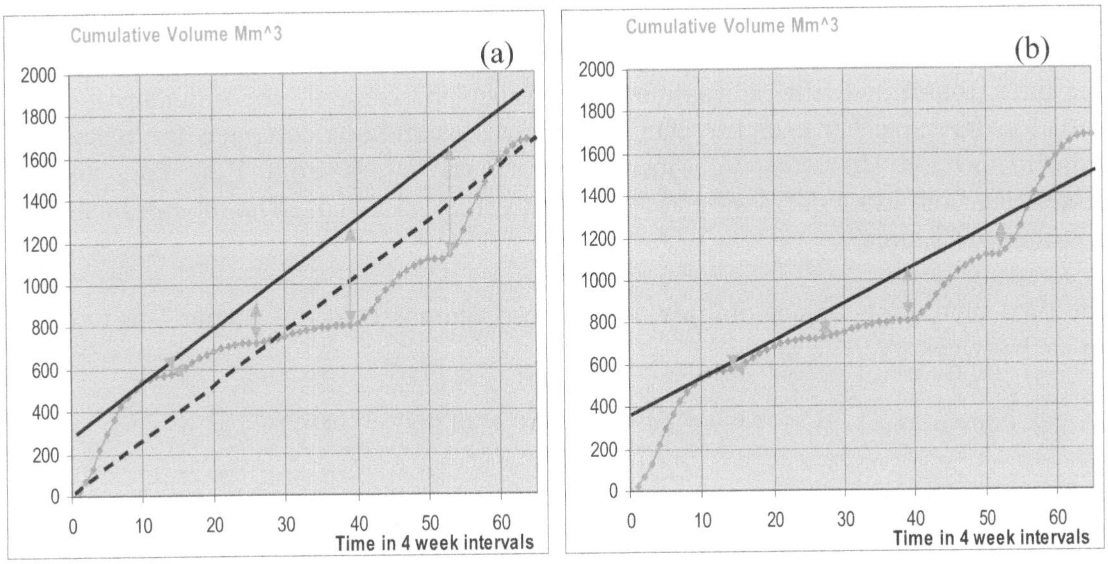

Figure 7.2.2: Mass curves for a 5 year hydrograph

Figure 7.2.2 illustrates a 5-year mass curve with 3 intervening drought years using an exaggerated vertical scale. It is then clear that there are 4 storage maxima associated with the difference between the tangential slope and the mass curve for each year. Figure 7.2.2(a) shows the 5-year available steady discharge and Figure 7.2.2(b) shows a lesser discharge. It is seen how the maximum necessary storage moves from the final year to the point of tangency as the yield is decreased. Note that for the reduced yield given in Figure 7.2.2(b), there is an excess of water that would be discharged over the reservoir spillway or released to the stream through outlet works. This excess occurs when the yield slope intersects the mass curve.

The success of this method depends on whether the reservoir has filled at the time of tangency of the required yield. This must be checked from the data. Many alternative design methods have been proposed (Shaw, 1994). In particular, if a long series of streamflow data is available, both the drought durations and their associated frequencies are best used in design.

7.2.4 Modules

Three modules are used for this project, namely “Hydrograph”, “ComputeTangent” and “Compute YieldCurves”. They are now described in detail.

“Hydrograph”

This module consists of 6 subroutines used for entering data into Columns A to F of the “Design” worksheet, and computing the design hydrographs. The first subroutine, “BoundaryCondition” has been described above, and ensures that Qp1> Qp2 and Qp3 >

Qp2. The second subroutine “InputInterval” completes Columns A and B in the “Design worksheet. Column A is the number of the time interval for each annual hydrograph, and Column B is the cumulative number.

The third, fourth and fifth subroutines are similar, and compute the ordinates for each annual hydrograph. For example, the “InputFlow1” subroutine computes the rising limb ordinate using a Do loop with increasing abcissa values from 1 to Tp1, and the descending limb using decreasing abcissa values from 13 to Tp1. The total ordinate data is written to Column C.

The final subroutine “InputColumns” completes Columns D to F. Column D is the mass for each 4-week interval, thus:

4-week interval mass = Q * 4 * 7 * 24 * 60 * 60 /1000000 = 2.4192*Q Mm^3

Column E is the cumulative mass for each interval, and column F is the slope of the cumulative mass curve for each interval of the first hydrograph. The latter is used in the next module for computing the yield tangent to the mass curve.

“ComputeTangent”

This module is the most complex, and consists of 3 subroutines of which the first 2 are similar. The first subroutine “Tangent” consists of 5 steps, as follows:

1. SpinButton 7 is set to its maximum scrolling value. This control is used for plotting the “Yield-Storage” curve and decrements in 50 Mm^3 stages of the cumulative volume. To ensure the decrement is in multiples of 50, the control is first set to the nearest multiple of 50 above the maximum value of the cumulative volume. (Note that this control is set before the “Tangent” subroutine is run, because the change in the control causes its associated control code to be run.)

2. The X (abcissa) and Y (ordinate) values, where the yield tangent intercepts the mass curve, are determined. The yield tangent slope is first determined (S) and then compared with each of the slopes for the mass curve intervals in column F, and interpolated to solve for X. Three other parameters are computed, namely the Y value at X (Y3), the Y value for X = 0 (Y4) and the Y value for maximum X (Y5). These values are written to Column G of the “Design” sheet, and used in later computation. Note that this step of the subroutine has an “Out of Limits” condition. The latter occurs when the tangent slope is greater than all of the interval slopes, and thus a tangent cannot be drawn to the mass curve. Such a condition can occur when Qp3 is set considerably higher than Qp1.

3. Two more “Out of limits” conditions are coded. These are necessary if the Y4 value becomes negative (for which the yield cannot be computed), and if the Y5 value exceeds 5000 (which is an arbitrary maximum for the project).

4. The maximum ordinate scale for the mass curve is computed. It is necessary to override the worksheet default maximum scale settings for the mass curve, as the tangent curve is also plotted on this chart. Since both the tangent curve and the storage indicator arrow are plotted in worksheet units, the maximum ordinate scale must be controlled.

5. The tangent is drawn. As indicated above, the maximum ordinate scale (M) is necessary to determine the worksheet units for drawing the tangent nodes.

The second subroutine "TangentVar" determines and plots the remaining "Yield-Storage" tangents according to the decrement of Spin button 7. The yield value is now taken from the Spin button value, and the remaining part of the tangent computation is the same as that in the "Tangent" subroutine. Since the chart ordinate scale has been set in the "Tangent" subroutine, it is not repeated. The same scale is used for plotting the decremented tangent.

Finally, the "DrawDifference" subroutine plots the storage indicator arrow on the mass curve. This is done in 4 steps, as follows:

1. The tangent and storage values are calculated for each abcissa of the mass curve and written to Columns I and J. The tangent values are simply calculated using the tangent slope and intercept (Y4). The mass curve values (column E) are then subtracted from these values to find the storage values.

2. The X value for each point of maximum storage is identified. This is done by finding the maximum storage (Ymax) using a worksheet function, then looping through the abcissa values to determine X for Ymax.

3. The Y value is calculated in worksheet units using the maximum scale (M) of the mass curve.

4. The storage is calculated in worksheet units.

"ComputeYieldCurves"

This module consists of 2 subroutines that compute the "Yield-Storage" curve in Columns K and L of the "Design" worksheet. The first "YieldCurveMax" subroutine computes the available, steady yield and enters these initial values as storage (C) in column K and yield (D) in Column L. The second "YieldCurveDown" subroutine computes the remainder of the "Yield-Storage" curve.

Note that the "YieldCurveDown" subroutine begins with the code:

```
Sheets("Run").SpinButton7.Max = Sheets("Run").SpinButton7.Value
```

This statement ensures that the yield curve can only decrease from its initial value.

Numerical check

The data chosen for this example are as follows:

Tp1 = 6; Tp2 = 19; Tp3 = 32 and Qp1 = 50; Qp2 = 10; Qp3 = 60

The check may be applied sequentially to each column in the "Design" sheet, as follows:

1. Column A has numbers 1 to 13 repeated 3 times
2. Column B has numbers 1 to 39 in series.
3. Column C can be checked for Qp1, Qp2 and Qp3, and the hydrograph chart will check the correctness of a synthetic triangle.
4. A multiplication formula can be used to check:
   Column D = 2.4192 * Column C
5. The cumulative sum in Column E can be simply checked by addition.
6. The interval slopes in Column F can be simply checked by subtraction.
7. Column G:
   Tangential slope = 1886.98/39 = 48.38
   Xtan = 9 + (51.84-48.38)/(51.84-34.56) = 9.20
   Y3 = 682.56 + 0.2(734.4-682.56) = 692.93
   Y4 = 692.93 – (9.2 * 48.38) = 247.80
   Y5 = 247.80 + (39 * 48.38) = 2134.77
8. Column I:
   I2 = 48.38 * 1 + 247.80 = 296.18
9. Column J:
   J2 = 296.18 – 20.16 = 276.02
10. Column K:
    K2 = maximum of Column J = 586.48
11. Column L:
    L2 = (48.38 * 10^6) / (4 * 7 * 24 * 60 * 60) = 0.41336 * 48.38
    = 20.00 m^3/s

EXERCISES

For all projects, change the given data and evaluate a numerical check for the new results.

Chapter 1. Visual Basic for Applications

1.1 The Visual Basic object model – "Draw.xls"

1.1.1 Draw the 2D Diagram on a new worksheet.
1.1.2 Draw a closed 2D object, and project it to 3D.
1.1.3 Record a Macro to rotate the object.
1.1.4 Add a DoLoop to the macro and rotate it to its maximum extent.
1.1.5 Add a Control button to activate the macro.

1.2 Objects and Procedures – "Objects.xls"

1.1.1 Change the number of random data for the Histogram.
1.1.2 Change the distribution function to $y = mx^3 - c$ in the Scatter sheet.
1.1.3 Write code to sort the rows from 0 to 1 for the array in the Surface sheet.
1.1.4 Write code to sort the columns from 0 to 1 for the array in the Surface sheet.

1.3 Numerical objects: estimating the Pi constant – "PIMC.xls"

1.3.1 Write code to list the object names in the "Run" worksheet.
1.3.2 Write the code to evaluate the definite integral $y = \int_0^{10} x\,dx$.
1.3.3 Write the code to evaluate the definite integral $y = \int_0^{5} (x^2+3)\,dx$.
1.3.4 Change the plotted data lines to points.

1.4 Kinematic objects: a manometer simulation – "Mano.xls"

1.4.1 Draw the object model for this project.
1.4.2 Draw the "Run" worksheet for this project.
1.4.3 Write a code to count the number of drawing objects on the "Run" worksheet.
1.4.4 Change the pressure range to ± 200 kPa.

1.5 Function objects: Evaporation and Transpiration – "Evap.xls"

1.5.1 Draw the object model for this project.
1.5.2 Find the daily evaporation for a Summer month (June) in the Atlantic Ocean (51°N) given n/nmax = 0.5, T = 20°C, u = 3 m/s, h = 0.5.
1.5.3 Find the daily evaporation for a Winter month (Jan) in the Gulf of Mexico (29°N) given n/nmax = 0.3, T = 15.2°C, u = 1 m/s, h = 0.9.
1.5.4 Find the daily evaporation for a Winter month (July) in the Pacific Ocean (10°S) given n/nmax = 1.0, T = 30°C, u = 5.5 m/s, h = 0.3.
1.5.5 Find the daily evaporation for a Summer month (June) in the Atlantic Ocean (51°N) given n = 2 hours/day, T = 20°C, u = 3 m/s, e = 5 mmHg.
1.5.6 Compare and explain the above 4 results.
1.5.7 Find the daily potential evapotranspiration for the above examples.

Chapter 2. Statistics and Geostatistics

2.1 **Linear and Multiple regression – "Regress.xls"**

2.1.1 Add +5 to rows 2,4,6… 20 and –5 to rows 3,5…21 of y linear data and explain the change in R^2.

2.1.2 Generate new random data using the "Objects.xls code", and regress.

2.1.3 Make a numerical check for a multiple regression.

2.1.4 Plot the 99% confidence limits for both regressions.

2.1.5 Add a CheckBox control to load and regress randomly generated data .

2.2 **Central Limit Theorem – "CLT.xls"**

2.2.1 Write code for n = 1, and explain results.

2.2.2 Write code for n = 16 and explain results (use loops for data).

2.2.3 Plot a set of results for each distribution, and for NN = 20,200,2000,20000.

2.2.4 Evaluate NN from the frequency and normal frequency curves.

2.2.5 Write code for a new probability distribution, say $f(x) = x^2$

2.2.6 Change the code to a 10 sided dice.

2.3 **Extreme value analysis – "EVA.xls"**

2.3.1 Change the MC sample limits to r = 100.

2.3.2 Add 3 readings to the data sample, and recompute.

2.3.3 Plot the 99% confidence limits.

2.3.4 Use the simplest formula, P = r/n, and explain differences.

2.3.5 Use the Weibull formula, P = r/ (n+1), and compare.

2.3.6 Find the limit of $t_{\alpha,v}$ for large n.

2.3.7 Add a control Check box, and show the trend of a MC added sample.

2.4 **Linear moments – "LMoment.xls"**

2.4.1 Using the Monte Carlo method, plot confidence limits for the L-Moment curves.

2.4.2 Show that E(X) is the subtended area of an idf for the linear and exponential functions in the "CLT.xls" code.

2.4.3. Write code for the 3 parameter Generalised Pareto distribution given by:

$$x(F) = \xi + \alpha\,[1 - (1 - F)^k] / k \quad \text{where } k = (1 - 3t_3) / (1 + t_3)\,;$$

$$\alpha = (1 + k)(2 + k) * l_2\,;\ \xi = l_1 - (2 + k) * l_2$$

2.4.4 Change the 5 quantiles to 10 deciles, and examine the curve changes.

2.4.5 Introduce extreme lower and upper outliers to a station, and check that they are followed by the L-Moment curves.

2.5 Ordinary and Universal Kriging – "OKrige.xls"

2.5.1 Add a 6th data point at 9, 6, 100.

2.5.2 Use an exponential semivariogram of the form g(h) = (1-exp(-h/R)).

2.5.3 With OK, R=4, spin K from 1 to 20; explain the changes in SE.

2.5.4 With OK, K=4, spin R from 1 to 20; explain the changes in SE.

2.5.5 Repeat the above 2 exercises for UK.

2.5.6 Why does the OK and UK code use an absolute value for SE; is it correct?

2.6 Block Kriging – "BKrige.xls"

2.6.1 Show the equivalence of the Point and Block Kriging codes.
2.6.2 Add a 6th data point to the code.
2.6.3 Use an exponential semivariogram of the form g(h) = (1-exp(-h/R)).
2.6.4 Use a linear semivariogram of the form g(h) = kh with the same sill.
2.6.5 Use ii = 900 in the functions module and discuss the new result.
2.6.6 Plot the convergence of solutions as ii is increased.

Chapter 3. Groundwater analysis

3.1 Groundwater levels – "GWL.xls"

3.1.1 Check the Linear Trend by spreadsheet.
3.1.2 Check the Quadratic Trend by spreadsheet.
3.1.3 Check the "goodness of fit" estimates by spreadsheet.
3.1.4 Use a 4th order polynomial with sill for both trends.
3.1.5 Add a third degree (cubic) trend surface module.
3.1.6 Explain the differences between sill values and the data variance.

3.2 Confined and unconfined aquifers – "AQF.xls"

3.2.1 Change the confined aquifer code to L = 2000 m and H = 30 m for a lake.
3.2.2 Change the unconfined well code to R = 1000 m. Discuss the effect on drawing scale.
3.2.3 Change the unconfined well code to a confined well.
3.2.4 Add a limestone soil with a hydraulic conductivity of 1 m/day.
3.2.5 Change the unconfined well, discharge units to m^2/day and discuss limitations.

3.3 Groundwater flow by finite difference – "GWFD.xls"

3.3.1 Summarise the iterations and residual standard deviations for a convergence criteria of 0.001, and compare the results with CC = 0.01.
3.3.2 Summarise the iterations and residual standard deviations with an initial head of zero. Discuss any differences with the existing settings.
3.3.3 Summarise the iterations and residual standard deviations with a convergence factor of 0.0001. Discuss any differences with the initial settings.
3.3.4 How does the relaxation factor influence the results (Try RF = 1).

Chapter 4. Precipitation

4.1 Design location of rainfall gauges – "UEden16.xls"

4.1.1 Move any of the 5 stations to see if the SE can be lowered.
4.1.2 Add a 6th station to the grid to minimise the SE.
4.1.3 Add a 7th station to the grid to minimise the SE.
4.1.4 Assume the rainfall trend passes exactly through the SSE, ESE lines and extrapolates to the catchment boundary, then Krige and discuss result.
4.1.5 Prepare a User Form to change the rainfall trend.

4.2 Mean catchment rainfall and error – "UEden349.xls"

4.2.1 Estimate the SE for a linear trend with a constant SV equal to the variance.
4.2.2 Do the same for a quadratic trend and explain the results for both trends.
4.2.3 Use Point Kriging for the data and explain the results.
4.2.4 Use nn = 1 for the data, and explain the results.
4.2.5 Run the code with nn > 60.
4.2.6 Provide a numerical check for ii = 16.
4.2.7 Write a User Form to adjust WR, nn and ii.
4.2.8 Run the code for a 2 km * 2 km grid.
4.2.9 Run the code for a 0.5 km * 0.5 km grid.
4.2.10 Test the residual SV for isotropicity.

4.3 Depth, duration and frequency curves – "DDF.xls"

4.3.1 Find the rainfall depth for a 100 year storm of 12 hours duration.
4.3.2 Find the rarity of a 100 mm rainfall depth with a 24 hour duration.
4.3.3 Find the duration of 50 mm rainfall depth with a 1/50 chance per year.
4.3.4 Find the annual maximum for a 24 hour duration, and compare it with AM-1d.
4.3.5 Extrapolate the single station DDF curves to T = 1000 years and discuss.
4.3.6 Evaluate and chart the confidence limits for the growth curves.

Chapter 5. Surface runoff

5.1 Round-nose horizontal broad-crested weir – "Weir.xls"

5.1.1 Calculate the velocity coefficient for H(T) = 0.5 m, b = 1 m, k = 0.1 mm. Use 2 methods and discuss.
5.1.2 Calculate the velocity coefficient for H(T) = 10 m, b = 100 m, k = 1 mm. Use 2 methods and discuss.
5.1.3 Add another control for k = 0.5 mm, and numerically check the results.
5.1.4 The dimensionless boundary layer thickness, δ/L, depends on the flow Reynolds number, UL/ν. U is the flow velocity, L the weir length and $\nu = 10^{-6}$ m^2/s is the coefficient of kinematic viscosity. Find the minimum values for Reynolds numbers used in this project.

5.2 Annual maximum flows – "QAM.xls"

5.2.1 The growth curve can be written: $x(F) = 1 + (\beta/k)\,[1 - ((1- F) / F)^k]$. Evaluate β as a function of k and t_2.
5.2.2 Evaluate the growth curve using β and discuss any simplifications.
5.2.3 Change the quantiles to T = 1.1, 2, 5, 50, 200 and rewrite the quantile subroutines.
5.2.4 Explain the extrapolation calculations used in the design hydrograph.

Chapter 6. Hydrologic cycle

6.1 Rainfall-runoff by the FEH method – "FEH.xls"

6.1.1 Add a control to reset the storm at any Scroll Bar value.
6.1.2 Change the area to X = 20 km and Y = 20 km and discuss the new design curve.
6.1.3 Change DPSBAR to 10 m/km and discuss the new design curve.
6.1.4 Change SAAR to 2000 mm and discuss the new design curve.
6.1.5 A 50% design summer storm profile has accumulated 50% of the storm depth. after 20% of the storm duration. Plot a design curve for a summer storm profile.

6.2 Rainfall-runoff by the SCS method – "SCS.xls"

6.2.1 For any particular storm profile, make a numerical check of the cumulative runoff.
6.2.2 For any particular storm, make a numerical check of the design peak flow for T = 75 years.
6.2.3 Add a hypothetical unit hydrograph to the analysis, and convolute.
6.2.4 For Area = 1 sq.ml and D = 24 h, find the minimum CN that can be used with this project, and explain.

Chapter 7. Reservoir design

7.1 Reservoir routing of floods – "ResRoute.xls"

7.1.1 Find the limits of the 6 variables for the condition of no flooding, when there is no spillway flow.
7.1.2 Find the limits of the 6 variables for the condition of no flooding when there is a spillway flow.
7.1.3 Use a second order polynomial for the storage curve trendline, and evaluate the increased accuracy.

7.2 Reservoir storage – "ResStore.xls"

7.2.1 Add to the code to calculate a drought of 2 years duration.
7.2.2 Plot a graph of Qp1 against Qp2 for fixed Qp2, when the code becomes "Out of limits", and explain.
7.2.3 Vary Qp2, and plot several Qp1, Qp2 graphs when the code becomes "Out of limits", and explain.
7.2.4 Why is the SpinButton7.maximum value (which is the next 50 Mm^3 multiple above the maximum mass curve value) never plotted.

BIBLIOGRAPHY

Computing

Bloch, S.C. (2000) "Excel for Engineers and Scientists", John Wiley and Sons, New York, USA.

Green, J., Bullen, S. and Martins, F. (1999) "Excel 2000 VBA Programmers Reference", Wrox Press, Birmingham, UK.

Halvorson, M. (2002) "Microsoft Visual Basic .NET step by step", Microsoft Press, Redmond, WA, US.

Heathcote, P.M. (2000) "Successful ICT Projects in Excel" Second Edition, Payne-Gallway Publishers Ltd, Ipswich, UK.

Jacobsen, R. (1999) "Excel 2000 Visual Basic for Applications", Microsoft Press, Redmond, WA, USA.

Krishnamoorthy, C.S. (1996) "Artificial intelligence and expert systems for engineers", CRC Press, Boca Raton, FL, US.

Liengme, B. V. (2000) "Guide to Microsoft Excel for Scientists and Engineers", 2nd Edition, Butterworth – Heinemann, 288 pages.

Smith, R. and Sussman, D. (1999) "Beginning Access 2000 VBA", Wrox Press, Birmingham, UK.

Walkenbach, J. (1999) "Microsoft Excel 2000 power programming with VBA", Hungry Minds Inc., USA.

Statistics

Chatfield, C. (1983) "Statistics for Technology", 3rd Edition, Chapman & Hall, UK, 384 pages.

Chatfield, C. (1996) "The analysis of time series", 5th. Ed., Chapman & Hall, UK, 300 pages.

Course Team (1984) "Unit 12 The Central Limit Theorem", Open University Press, Milton Keynes, UK.

Gumbel, E.J. (1958) "Statistics of extremes", Columbia University Press, 375 pages.

Hosking, J.R.M. and Wallis, J.R. (1997) "Regional frequency analysis", Cambridge University Press, Cambridge, UK.

Kite, G.W. (1977) "Frequency and risk analyses in hydrology", Water Resources Publications, Fort Collins, CO, USA.

Montgomery, D.C., Runger, G.C. and Hubele, N.F. (2001) "Engineering statistics", 2nd Edition, John Wiley and Sons, New York, USA.

Mooney, C.Z. (1996) "Monte Carlo simulation", Sage Publications Inc., Iowa City, USA.

Stuart A. and Ord, J.K. (1987) "Kendall's advanced theory of statistics", Vol.1, Charles Griffin & Co., London, UK.

Geostatistics

Bras, R.L. and Rodriguez-Iturbe, I. (1993) "Random functions and Hydrology", Dover Publications, New York.

Clark, I and Harper, W.V. (2000) "Practical Geostatistics 2000", Ecosse North America LLC, USA.

Davis, J.C. (1986) "Statistics and data analysis in Geology", 2nd Edition, John Wiley & Sons, New York, USA.
Hohn, M.E. (1988) "Geostatistics and petroleum geology", Van Nostrand Reinhold, New York, USA.
Isaaks, E.H. and Mohan Srivastava, R. (1989) "Applied Geostatistics", Oxford University Press, New York, USA.
Kitanidis, P.K. (1997) "Introduction to Geostatistics", Cambridge University Press, Stanford-Cambridge Program, Cambridge, England.
Krige, D.J. (1966) "Two-dimensional weighted moving average surfaces for ore evaluation" Proc. Symposium on Mathematical Statistics and Computer Applications in Ore Valuation, Mar 7, Jo. South African Inst. Mining and Metallurgy, Johannesburg, pages 13-38.
Matheron, G. (1963) "Principles of Geostatistics", Economic Geology, Vol. 58, pages 1246-1266.
Olea, R.A. (1975) "Optimum mapping techniques using regionalised variable theory", Kansas Geological Survey Series on Spatial Analysis , No. 2, 137 pages.
Webster, R and Oliver, M.A. (2001) "Geostatistics for Environmental Scientists", John Wiley & Sons, West Sussex, England.
Wellmer, F-W (1998) "Statistical evaluations in exploration for mineral deposits", Springer-Verlag, Berlin, Germany.

Hydrology

Ackers, P., White, W.R., Perkins, J.A. and Harrison, A.J.M. (1978) "Weirs and Flumes for Flow Measurement", John Wiley & Sons, Chichester, UK.
Bear, J. and Verrujit, A. (1990) "Modeling Groundwater Flow and Pollution", D. Reidel Publishing company, Dordrecht, Holland.
Bos, M.G. (1978) "Discharge Measurement Structures", 2nd Edition, Intenational Institute for Land Reclamation and Improvement, Wageningen, The Netherlands.
British Standards Institution, Part 4F (1990) "Methods of measurement of liquid flow in open channels. Weirs and flumes. Round nose horizontal broad-crested weirs", BS 3680, BSI, London, UK.
Dwyer, I.J. and Reed, D.W. (1995) "Allowance for discretisation in hydrological and environmental risk estimation (ADHERE)", IH report No. 123, Institute of Hydrology, Wallingford, UK.
Faulkner, D. (1999) "Flood Estimation Handbook, Volume 2: Rainfall Frequency estimation", Institute of Hydrology, Wallingford, UK.
Faulkner, D. (1999) "Flood Estimation Handbook, Volume 3: Statistical procedures for flood frequency estimation", Institute of Hydrology, Wallingford, UK.
Hoggan, D.L. (1996) "Computer-assisted floodplain hydrology and hydraulics", 2nd Edition, McGraw Hill, New York, USA.
Houghton-Carr, H. (1999) "Flood Estimation Handbook, Volume 4: Restatement and application of the Flood Studies Report", Institute of Hydrology, Wallingford, UK.
Houghton-Carr, H.A. and Boorman, D.B. (1991) "A national archive of flood event data for the UK", Proc. BHS Third National Hydrology Symposium, Southampton, British Hydrological Society, Wallingford, UK.

Linsley, R.K. and Franzini, J.B. (1991) "Water Resources Engineering", 4th Edition, McGraw Hill, New York, USA.

McCullough, J.S.G. (1965) "Tables for the rapid computation of the Penman estimate of evaporation", Jo. E. African Agric. Forest., Vol. XXX(3), pages 286-295.

Meteorological Office (1981) "The Meteorological Office Rainfall and Evaporation Calculation System, MORECS", Hydrological Memorandum No. 45, UK.

Penman, H.L. (1948) "Natural evaporation from open water, bare soil and grass", Proceedings of the Royal Society of London, Vol. 193, pages 120 – 145.

Ministry of Agriculture, Fisheries and Food (1967) "Potential Transpiration", MAFF, Technical Bulletin 16, HMSO, UK.

Roberson, J.A., Cassidy, J.J. and Chaudry, M.H. (1998) "Hydraulic Engineering", 2nd Edition, John Wiley & Sons, New York, USA.

Sharp, J.J. and Sawden, P. (1984) "BASIC hydrology", Butterworths, London, UK.

Shaw, E.M. (1994) "Hydrology in Practice", 3rd Edition, Chapman and Hall, London, UK.

Stewart, E.J. (1989) "Areal reduction factors for design storm construction", International Association of Hydrological Sciences, Pub. No. 181, pages 31-40, Wallingford, UK.

U.S. Natural Resources Conservation Service (Formerly U.S. Soil Conservation Service, SCS). (1964) "National Engineering Handbook", Dept. of Agriculture, Washington, D.C.

U.S. Weather Bureau (1961) "Rainfall Frequency Atlas of the U.S", Technical Paper No. 40, Dept. of Commerce, Washington, D.C.

Wang, H.F. and Anderson, M.P. (1982) "Introduction to Groundwater Modeling", Academic Press, San Diego, US.

Wilson, E.M. (1990) "Engineering Hydrology", 4th. Edition, Macmillan Education Ltd, Basingstoke, UK.

Wilson, E.M. (1990) "Engineering Hydrology, Solutions to Problems", 4th. Edition, Macmillan Education Ltd, Basingstoke, UK.

Index

IWA Publishing's authorised EU representative for General Product Safety Regulations is Diane D'Arras, 15 rue Duret, 75116 Paris, France, e-mail: safety@iwap.co.uk.

Printed and bound by CPI Group (UK) Ltd, Croydon, CR0 4YY
12/05/2026
02108886-0016

IWA Publishing's authorised EU representative for General Product Safety Regulations is Diane D'Arras, 15 rue Duret, 75116 Paris, France, e-mail: safety@iwap.co.uk.

Printed and bound by CPI Group (UK) Ltd, Croydon, CR0 4YY
12/05/2026
02108892-0017